# ROAD MAINTENANCE AND REGRAVELLING USING LABOUR-BASED METHODS

## HANDBOOK

THE ROMAR TRAINING MANUALS FOR CONTRACTORS

# Road maintenance and regravelling (ROMAR) using labour-based methods

# HANDBOOK

*Prepared for the International Labour Office*
*by*

CLAES-AXEL ANDERSSON
ANDREAS BEUSCH
and
DEREK MILES

INTERMEDIATE TECHNOLOGY PUBLICATIONS 1996

Intermediate Technology Publications Ltd
103–105 Southampton Row, London WC1B 4HH, UK



A CIP record of this book is available from
the British Library

ISBN 1 85339 348 7

Typeset by Dorwyn Ltd, Rowlands Castle, Hants
Printed by SRP, Exeter, UK

---

# CONTENTS

# PREFACE

ROMAR stands for labour-based *Ro*utine *Ma*intenance and *R*egravelling, and extends the *Improve Your Construction Business (IYCB)* approach to small contractor development into this technical area. Readers who are familiar with the three IYCB handbooks and workbooks[1] will feel at home with the ROMAR handbook and workbook. As with IYCB, the ROMAR handbook provides ideas and information and the workbook gives readers a chance to look at their business in a disciplined way, and decide on action plans to make it more competitive and successful.

As more countries appreciate the social and economic benefits that can come from applying labour-based construction techniques, together with the productivity gains that can be achieved by entrusting this work to efficient private contractors, the market for ROMAR activities can be expected to grow. The ROMAR books will be particularly useful for experienced general contractors who are new to routine road maintenance and regravelling, and who need to judge whether their own resources and skills will enable them to turn this into an attractive business opportunity.

The ROMAR handbook and workbook are each divided into two parts, the first covering principles and the second dealing with the practice of labour-based road construction and maintenance. The 'route map' in the section on 'How to use your ROMAR books' will help you to find your way around, and concentrate on those chapters which are most relevant for your own circumstances. The books can be used in the context of an integrated training programme including opportunities to work on pilot contracts under supervision, but they have also been designed in a format suitable for self-study.

This book was prepared and edited within the Employment-Intensive Works Programme of the ILO based in its Development Policies Branch.

Claes-Axel Andersson
Andreas Beusch
Derek Miles

---

[1] *Improve Your Construction Business series. No. 1: Pricing and bidding. No. 2: Site management.* No. 3 Business management. ILO Geneva, 1994/6.

# THE AUTHORS

**Claes-Axel Andersson** is an independent consultant working with construction management issues, including small-scale contractor development. Until March 1996 he was an expert with the Employment-Intensive Works Programme of the ILO, based in its Development Policies Branch, also being responsible for ILO's Construction Management Programme including its Improve Your Construction Business activities. Mr. Andersson is a professionally qualified civil engineer with extensive experience in project management and building design.

**Andreas Beusch** is an independent consultant, specializing in appropriate road-work technology and training. He has over 15 years of experience, practical and managerial, of the development of labour-based construction methods in several developing countries. Mr. Beusch is a professionally qualified site engineer who started his career with his own planning and contracting firm.

**Derek Miles** is Director of the Institute of Development Engineering at Loughborough University, United Kingdom and is also Director of Overseas Activities in its Department of Civil and Building Engineering. He is a Fellow of the Institution of Civil Engineers and the Institute of Management and has more than 20 years' experience in the development of national construction industries. He directed the ILO Construction Management Programme during the period 1986–94.

# ACKNOWLEDGEMENTS

The Improve Your Construction Business (IYCB) approach to development of small-scale contractors was initiated through a pilot project in Ghana financed by the Government of the Netherlands. The IYCB concept has since been successfully introduced in other countries in Asia and Africa. The IYCB material, developed according to a modular concept, was already, at the conception stage, foreseen to be supplemented with training material covering technical and specific management issues for different construction sub-sectors.

In connection with a major World Bank infrastructure programme, the Government of Lesotho asked the ILO to undertake a contractor development project in the labour-based road sector. A major initial activity of this project was the production of a Road Maintenance and Regravelling (ROMAR) package expanding the IYCB concept into the labour-based road sector.

The ROMAR training sessions of the project were undertaken in close collaboration with the Labour Construction Unit (LCU) of the Lesotho Ministry of Works. The authors would like to recognize the dedication and enthusiasm of the LCU training staff that provided extensive assistance to the two rounds of contractor training. We would also like to recognize the two rounds of training provided comprehensive feed-back from trainees and trainers, which was incorporated when finalizing the manuscript.

# HOW TO USE YOUR ROMAR BOOKS

This book is written for you – the owner or manager of a small construction business. Together with the three basic IYCB handbooks and workbooks, the ROMAR books provide you with both management and technical advice that you need in order to make a success of routine maintenance and regravelling as a business activity. As with IYCB, the ROMAR handbook and workbook are best read together. We suggest you first read the chapter in the handbook, and then work through the examples in the corresponding chapter of the workbook.

## The handbook

The handbook, like the workbook, is divided into ten chapters. Part A, the first five chapters, deals with principles. The remainder of the book helps you to put these principles into practice.

Chapters 1 and 2 will be of most interest to those contractors who are new to the roads business. Chapter 1 summarizes the technical considerations, and also explains the most important engineering standards for earth and gravel roads. Chapter 2 provides an introduction to the choice of appropriate road construction and maintenance technology, and is written around a case study in which a new contractor looks at the business aspects of road works and decides on the type of equipment which will be required in order to carry out the various activities with a minimum capital outlay.

Chapter 3 introduces the reader to basic soil mechanics, including soil identification procedures, simple field tests and the principles of compaction. Chapter 4 deals with equipment, vehicles and tools, including repair and maintenance requirements; planning and reporting systems, service schedules and descriptions and basic specifications for common hand tools to assist in procurement. Chapter 5 completes the section on principles with a description of the major operations in a typical labour-based road project, and concludes with a case study drawn from one of the most successful labour-based programmes in Africa.

Once you are sure you have understood the principles set out in Part A, you will be ready to move on to Part B. Chapter 6 is a general chapter which explains how roads can deteriorate, the three main road maintenance systems, and the way in which maintenance contracts should be managed. If you are interested in routine maintenance you will then proceed to Chapter 7, which describes how to plan and carry out the 12 main routine maintenance activities. Chapter 8 has been written for those contractors who intend to specialize in regravelling, and describes how to plan and carry out these project-based operations. Even if you intend to specialize in only one of the two types of road maintenance, we recommend that you should look through both chapters before concentrating on the one in your own speciality area.

The book concludes with two chapters on ROMAR as a business activity, to supplement the topics dealt with in more general ways in the IYCB series. While the IYCB books relevant to labour-based road works go through pricing and bidding in detail, Chapter 9 in this book deals with special considerations relevant for road maintenance contracts, and shows how to build up detailed prices for a bill of quantities. Since labour-based projects inevitably mean that you have to cope with large workforces, Chapter 10 provides practical advice on how to manage people, including team building, communication, training, motivation, incentives, and how to ensure good discipline and morale. The book concludes with a list of reference literature and a summary of ILO Labour Standards (see Annexe).

## The workbook

The workbook enables you to test your understanding of ROMAR, and decide whether you have the knowledge, experience and resources to make a success of this business activity.

In each chapter of the workbook there is a list of simple questions to which you answer 'yes' or 'no'. The answers will tell you about the strengths and weaknesses of your existing business, and whether or not you should take the risk of venturing into one or both of the ROMAR activities (routine maintenance and/or regravelling).

If you have decided to go ahead, but still need to improve your knowledge in certain areas, you can then turn back to the appropriate section in the handbook and try again. Even when you have established yourself as a successful ROMAR contractor you will probably find it useful to go back to your handbook and workbook from time to time to see if there are ways to further improve your performance.

Please note that calculations are normally based on a number of assumptions, thus the answers to most calculations in both the handbook and the workbook are rounded to the nearest whole number or round figure, or to the nearest ten or hundred in the case of large numbers. This is done to avoid giving the impression that the end-result is more accurate than it actually is. Students using calculators should expect similar rather than identical answers in such estimates.

## Where to start

We recommend that you start by reading quickly through the whole of the handbook. Then you can go back over it more slowly, concentrating on the chapters which deal with topics which are new to you.

The following 'route map' (overleaf) will help you to find your way around the handbook.

> Note: As this book is intended for use in many different countries, we have used the term 'NU' in the examples to represent an imaginary National Unit of currency, and 'NS' to stand for imaginary National Standards.

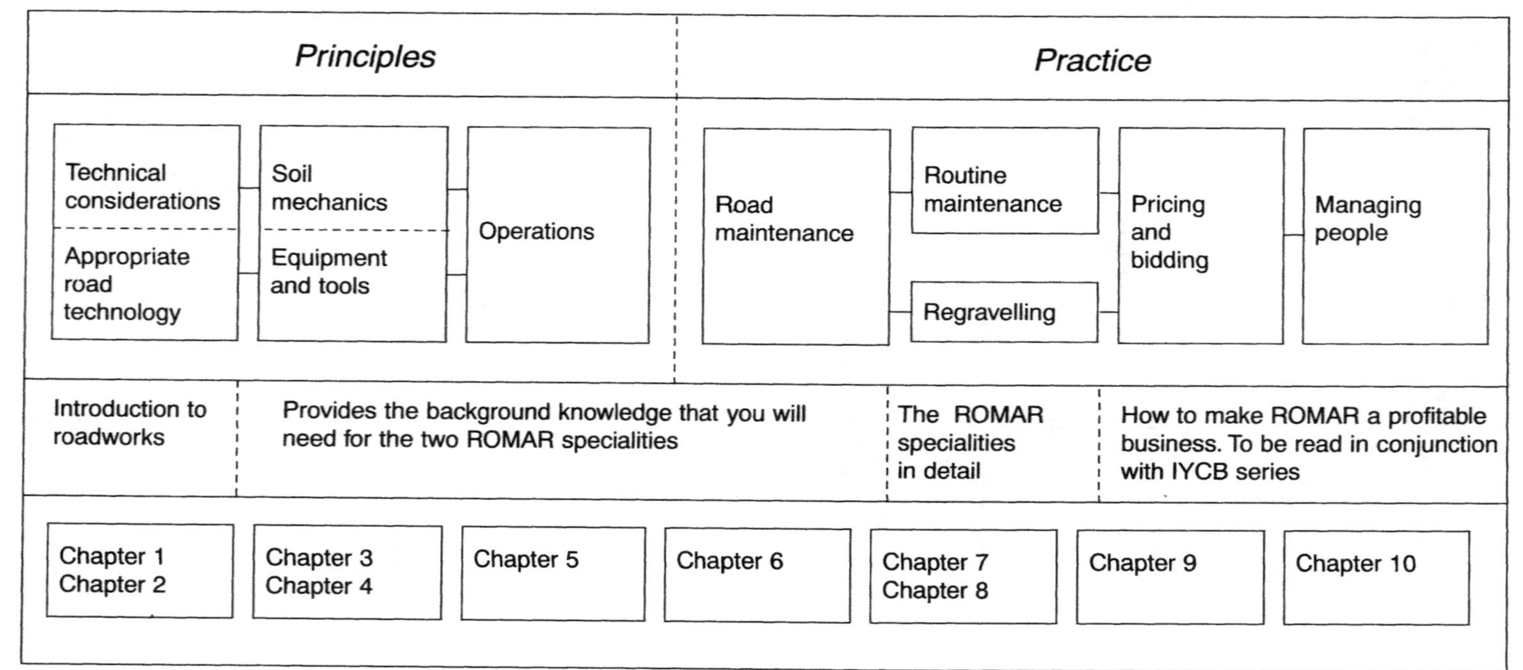

Principles
Practice
Technical considerations
Appropriate road technology
Soil mechanics
Equipment and tools
Operations
Road maintenance
Routine maintenance
Regravelling
Pricing and bidding
Managing people
Introduction to roadworks
Provides the background knowledge that you will need for the two ROMAR specialities
The ROMAR specialities in detail
How to make ROMAR a profitable business. To be read in conjunction with IYCB series
Chapter 1 Chapter 2
Chapter 3 Chapter 4
Chapter 5
Chapter 6
Chapter 7 Chapter 8
Chapter 9
Chapter 10

# SECTION A

# PRINCIPLES

# CHAPTER 1: ROADS: THEIR PURPOSE, TERMINOLOGY AND STANDARDS

## Learning Objectives

After you have completed this chapter you should know:

- the main types of roads and how they contribute to meeting people's needs for transportation
- the basic road terminology in relation to labour-based road works
- the most common engineering road standards in relation to labour-based road works.

## Introduction

You are interested in becoming a contractor for regravelling and/or routine maintenance of roads using mainly labour-based methods. As a contractor your foremost interest is to find good business opportunities for your company. Labour-based road works can offer such an opportunity for contractors with knowledge, experience and the right attitude. After reading the Handbook and testing your knowledge in the Workbook you will be in a better position to judge whether you can succeed in this specific area of work.

In this first chapter we will provide you with background information about the product you will deal with as a road work contractor: the roads themselves. We shall briefly discuss their general purpose, the relevant terminology and the most common standards. This will enable you to gain a basic knowledge about roads so that you can communicate with the client and understand the purpose of the tasks that are required in a typical road contract.

# The Purpose of Roads

Transport is essential for the economy and the social welfare of a country. Roads are a vital part of the transportation system, and a good road network will:

- enable goods and services to be transported efficiently
- enable buses, taxis and private cars to move people from their homes to their places of work, to public or social facilities, and to market places
- allow freedom of movement for the people of a country to visit areas which may otherwise not be reached
- allow the government administration to reach people where they live in order to provide them with the necessary services
- promote local industry and trade.

# Construction Standards

There are three principal types of roads according to construction standards:

- paved roads
- gravel roads
- earth roads

*Paved roads*

Paved roads are roads which have a homogenous and strengthened carriageway surface which usually consists of bitumen or concrete. Other, less used, alternative pavements are concrete blocks, clay bricks or hand-packed stones. Most of the major roads in a country are usually of bitumen standard, which allows the efficient transportation of goods and people.

*Gravel Roads*

Gravel roads are very common in developing countries. The fact that these roads have a gravel surface means that they are all-weather roads which can be used all the year round.

*Earth Roads*

Earth roads are also very common in developing countries and often make up the majority of the road network. They are cheap, but have the disadvantage that they usually have a low technical standard and can deteriorate rapidly in high rainfall areas or where heavy traffic passes. Earth roads are, in most cases, rural access roads and are very useful for the local population. Animal-drawn carts, bicycles and pedestrians are the major users of these roads.

## Classification

Besides being divided according to the types of construction standards, as described above, roads can also be divided into classes of importance or function within a country's road network. Classification systems vary, but usually include:

- international trunk roads: main roads which cross borders and connect neighbouring countries
- national trunk roads: roads which connect major parts of the country, often between the country's main cities
- regional roads: roads which connect provinces or regions in a country and connect or provide access to their main towns or villages
- minor roads: roads which connect villages and market centres to the regional and national roads
- rural roads: roads which connect the rural areas to market centres, villages and/or the higher classified road network
- special purpose roads: roads which serve a special purpose, such as security roads, sugar roads, wheat roads, tea roads or forestry roads.

## Road Works Terminology

If you are in the business of road construction, rehabilitation or maintenance, you must know the common technical terms used to describe a road on drawings and in specifications because:

- you cannot estimate realistically without understanding the contract documents
- you need to use these terms accurately when discussing work to be done with your clients
- you need to be able to explain these terms to your site supervisory staff in a language they can understand.

This section provides you with definitions of the most important road terminology. As a small-scale contractor you will mainly deal with gravel and earth roads and we will therefore explain terms applicable to these types of roads.

## CROSS-SECTION TERMS

To make it easier to understand, we have provided a cross-section of a typical gravel road and indicated the main road components in Figure 1.1. The numbers refer to those in the figure.

*1. Formation Width*
Full width of road, including drains and embankments.

*2. Roadway*
Width of road, including shoulders.

*3. Carriageway*
Paved width of the road, available for traffic.

*4. Shoulders*
Paved or unpaved width of road next to the edge of the carriageway adjacent to the ditch or embankment slope.

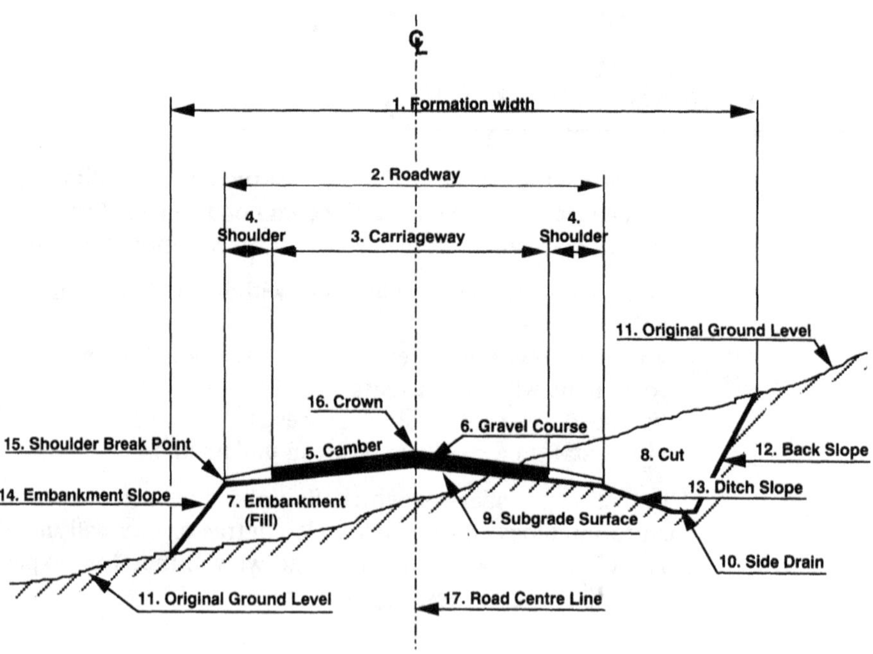


Figure 1.1 *Cross-section terms*

*5. Camber*
A cambered road has a cross-section like the roof on a house, to drain the rainwater away from the carriageway to the side drains.

*6. Gravel course*
A layer of compacted gravel which forms the surface (or pavement) of the carriageway.

*7. Embankment*
Compacted earth fill below the roadway.

*8. Cut*
Excavation in the natural ground on the hill side of the road usually with graded slopes. The material dug out is used to fill the embankment on the valley side of the road.

*9. Subgrade surface*
Upper layer of the soil (natural material) supporting the roadway including embankment slopes.

*10. Side drain*
The side drains run along the road and collect the water from the carriageway and adjoining land and transport it to a convenient point of disposal.

*11. Original ground level*
The natural surface of the cross-section prior to construction.

*12. Back slope*
The outer slope of the side drain with an appropriate angle to prevent the soil from sliding to the ditch.

*13. Ditch slope*
Inside slope from the shoulder to the side drain.

*14. Embankment slope*
Natural material slope on embankment.

*15. Shoulder break point*
The junction of the carriageway shoulder with the drainage ditch.

*16. Crown*
Peak or highest point of the camber.

*17. Road centre line*
Line running along the centre of the road (important in surveying and setting out the road alignment). Chainage is a term frequently used for describing distances measured along the centreline of a road.

## DRAINAGE TERMS

Good drainage is vital if roads are going to survive heavy rains. The following definitions will help you to understand the discussions on drainage in later chapters. Figure 1.2 shows you where the major drainage elements are located. The numbers refer to those in the figure.

*1. Side drains*

The side drains run along the road and collect the water from the carriageway and adjoining land and transport it to a convenient point of disposal.

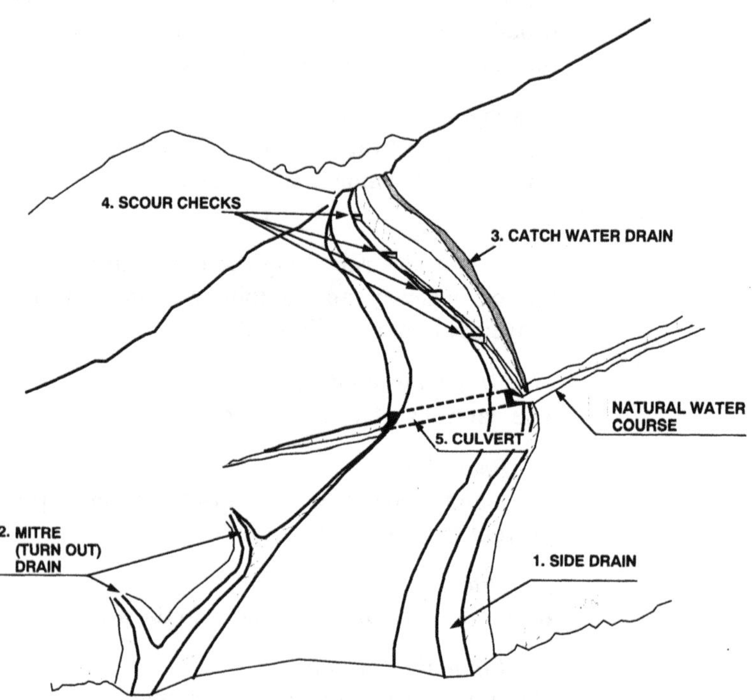


Figure 1.2 *Drainage elements*

*2. Mitre drains*

Mitre drains (or turn-out drains) lead the water out of the side drains and safely disperse it on adjoining land. Mitre drains should be provided as often as possible so that the accumulated water volume in each drain is not too high and does not cause erosion to the adjoining land.

*3. Catch water drains*
Where the road is situated on a hillside a significant amount of rainwater may flow down the hill towards the road. This may cause damage to the cut face (backslope) of the road and may even cause land slides. Catch water drains intercept or 'catch' surface water flowing towards the road from adjacent land, and lead it away.

*4. Scour checks*
Scour checks prevent erosion in side drains on steep gradients by slowing down the water (steps). Scour checks are usually built using locally-available material, such as stones or wooden sticks. (Construction details are explained in Chapter 5).

*5. Culvert*
The culvert is a transverse drain built under the road and its function is to lead water from the upper, uphill side of the road to the lower, valley side. In tropical countries with high rainfalls three or four culverts are required per kilometre. Culvert rings are usually made of concrete or prefabricated corrugated steel rings. (See Figure 1.3.)

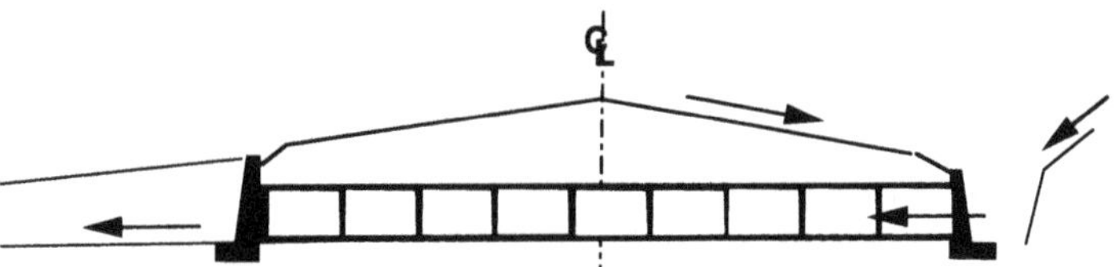

Figure 1.3 *Culvert*

## MAJOR STRUCTURES

*Bridges*
Bridges are designed to allow natural water courses like rivers to pass below the roadway elevation.

Figure 1.4 *Bridge*

*Drift or splash*

Bridges are expensive, so an alternative is to construct a drift – a low-level crossing which constant or seasonal water collected from the ditches and/or natural water courses can flow over. Although drifts are less convenient than bridges for road users, and can become unusable at times of heavy rains, their low cost offers a great attraction when funds are short. (See Figure 1.5.)

Figure 1.5 *Drift/splash*

*Vented ford (or vented drift)*

The vented ford is an intermediate solution between a drift and a bridge. It is a medium-level stream or river crossing through which the normal flow of water can pass, but which is designed to be over-topped during periods of heavy rainfall.

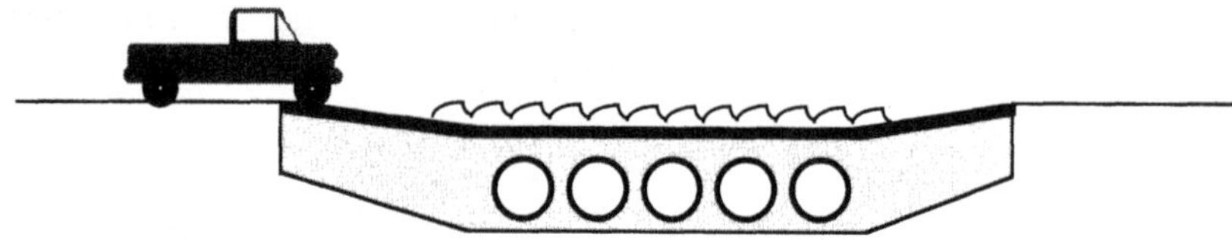

Figure 1.6 *Vented ford*

## GENERAL ROAD-BUILDING TERMS

You should also be familiar with the following general terms.

*Construction*

The process by which a road is actually built according to established design standards and construction plans.

*Rehabilitation*

Activities which restore a road to the original recommended design standards. This usually means improving an existing deteriorated road and restoring its geometry.

*Upgrading*

The process by which the standard of an existing road is improved to allow safe use by a greater volume of traffic than it was originally designed for.

*Maintenance*

The work which is required to retain the original standard of the road. (The detailed maintenance terms are explained in Chapter 6).

# Principal Standards for Earth and Gravel Roads

This Section explains the most important standards for earth and gravel roads, taking into consideration that they are constructed and maintained using labour-based methods and basic equipment. These standards, drawing upon the experience of numerous projects involving labour-based road technology in Africa and Asia, are generally applied in projects of this kind.

Specific project standards are usually developed by the planners of the project. You, as a contractor, will therefore get the road design, including all the construction or maintenance standards, from your client as part of the contract documentation. In order to properly prepare a bid for the work and later on to implement the contract, you must be able to understand and interpret these standards.

*Carriageway width*

The overall aim of the road designer is to provide safe all-weather passage for the users at minimum construction cost. The carriageway width has a profound effect on construction costs, and standard widths are set by road authorities taking into account the road alignment and the volume and type of traffic using the road. It is essential that there is provision for vehicles to pass each other. In mountainous terrain, where the road cross-section must be reduced, passing places have to be provided.

A typical maximum carriageway width for roads to be constructed using labour-based methods is 5.50m and a typical minimum carriageway width is 3.50m.

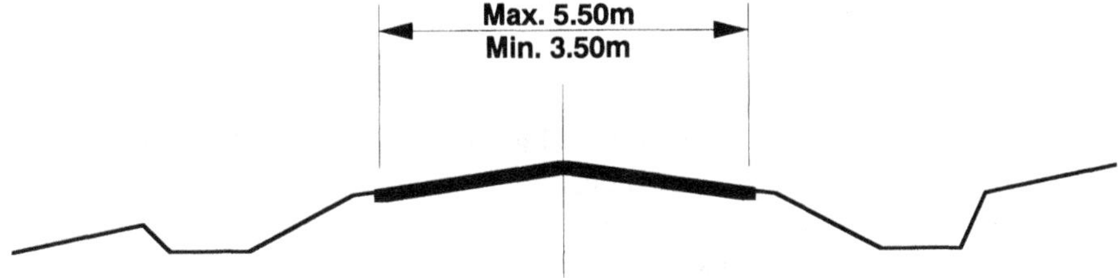


Figure 1.7 *Carriageway width*

*Camber*

The purpose of the camber is to shed water as rapidly as possible from the roadway and allow it to drain into the roadside ditches. Experience has shown that the camber of gravel surfaced roads should be minimum 5% to be effective, but not in excess of 8% as that can cause erosion of the pavement. In addition, a camber of 8% or more reduces the riding comfort. The camber must be included during initial formation of the roadway to ensure the economic and efficient use of the surfacing material.

Where the road formation cannot be compacted during construction the camber has to be increased to approximately 10%. After the formation material has settled the final camber will be between 5% and 8%.

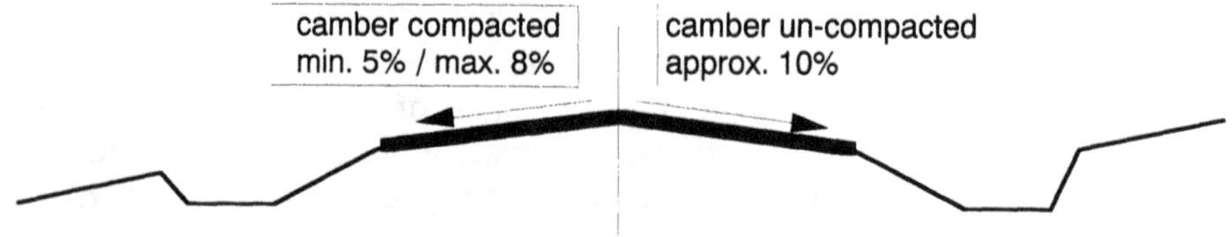


Figure 1.8 *Camber*

*Side drains*

The conventional shape of a side drain is a 'V', as this is the only shape a grader can cut. Roads constructed and maintained using labour-based methods require a 'U' shape (or trapezoidal shape) drain. It is much easier to dig and clean a U-shaped drain with hand tools, like hoes and shovels, and the risk of erosion is lower. The minimum width of a side drain is 40cm.

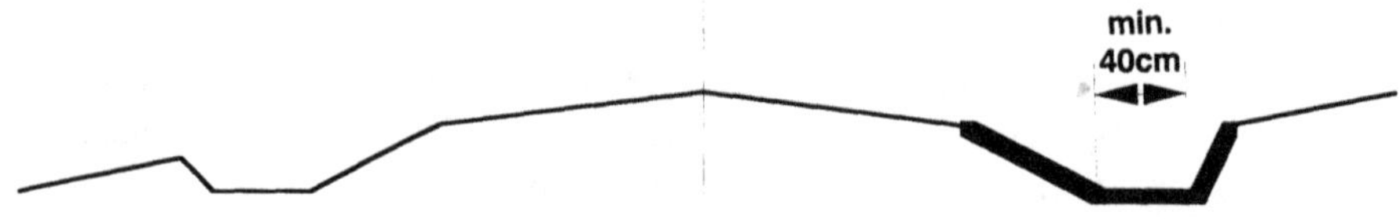


Figure 1.9 *Side drain*

*Gradients*

A minimum longitudinal gradient of 2% (gradient along centre line) of the road is desirable for adequate drainage. Where this minimum gradient cannot be achieved severe drainage problems may result, as it is very difficult to lead the drain water away from the road. The only solution may be very long mitre drains. That, however, often causes trouble because the flow of water will not be sufficient to take away all the silt in the run-off water. In these cases it will be necessary to allow for the cost of regular silt removal as part of the routine maintenance activities.

At the other extreme the desirable maximum gradient is about 8% to 10% on gravel and earth roads and such steep gradients should not exceed a length of 100m. Most loaded vehicles would not be able to climb gradients steeper than 8–10% on a gravel or earth surface. (See Figure 1.10.)

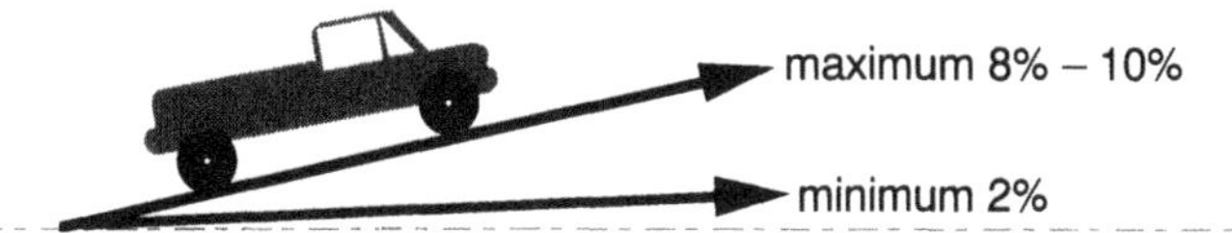


Figure 1.10 *Gradient*

*Slopes*
Where cuts have to be made, unprotected slopes should be constructed with an inclination that will not allow the soil to slip or erode. The appropriate inclination (natural material slope) differs according to the type of soil. The inclination of slopes at the side of roads will be set out by the client in the work specifications.

Other standards, such as the minimum sight distance, minimum curve radius and so on, can differ from project to project. Standards for gravel will be discussed in Chapter 3 (Soils). Normally standards are included in the contract specifications, but remember that it is your responsibility to check with the client about the exact standards whenever you are in doubt.

# CHAPTER 2: APPROPRIATE ROAD CONSTRUCTION AND MAINTENANCE TECHNOLOGY

## Learning Objective

After you have completed this chapter you should know:

- how to make your best choice for road construction and maintenance technology.

## Introduction

In many developing countries the methods used for road construction, rehabilitation and maintenance are similar to the technologies established in the industrialized world. In industrialized countries labour costs are generally high, while plant and equipment can be bought and operated relatively cheaply. So it often makes sense to replace costly labour with cheaper equipment. In developing countries, on the other hand, unemployment is high, labour is relatively cheap, but equipment is expensive to buy, operate and maintain. So the best, or most appropriate, technology for a developing country can be very different from that which is most appropriate for an industrialized country.

In this chapter we shall discuss what choice of technology you, as a contractor for road-works, should make to suit the needs of your business (to make a good profit) as well as the needs of your country (to create employment). To illustrate what the ideas presented here mean to a small-scale entrepreneur in practice, a typical contractor and his company are introduced step-by-step throughout this chapter.

Additional information on the subject of the choice of technology is provided in IYCB Handbook 2 'Site Management', Chapter 1.

*Example:*
*As a young building technician you decided two years ago to start your own business as a building contractor. You started your business with hope and courage, but little money. Your family*

*lent you a little capital which allowed you to buy a second-hand concrete mixer, some basic hand tools and an old pick-up.*

*As a first job you were given a contract to rehabilitate your uncle's house. Later on you were awarded a small contract from the town council to lay a sewer, and as a result of your efficient work you were awarded more small contracts of various types. As you are committed to a career in the industry you have joined the local contractors' association, and you have been registered as a contractor class 'D'.*

*Six months ago you decided to get a loan from a bank to buy two second-hand tipper trucks for 45 000 NU. You bought them rather cheaply but had to spend quite a lot to rehabilitate them, which cost you an additional 33 500 NU. One of your brothers is a qualified mechanic and he carried out the overhaul quickly and well. He is now fully employed by you, as the two trucks and the pick-up keep him busy. The trucks are being used to transport your own material, and also for haulage on contract for other clients.*

## The Market for Small-scale Road Work Contractors

As a small-scale contractor you have to work within a business environment that you cannot change or influence. The economies of many developing countries have deteriorated and foreign exchange is scarce. Interest rates on bank loans are extremely high and the costs of imported road construction and maintenance equipment have increased, while local labour costs have declined in hard currency terms.

*Example:*
*Two months ago you were informed by your contractors' association of a new programme that will construct, rehabilitate and maintain access and feeder roads in the area, and which intends to use small-scale contractors. You are interested in the new programme as you think that it could offer reliable and profitable work. With your family members and 'advisers' you immediately start to make plans. You talk to an expatriate engineer working for a large international contractor building the first motorway in your country and are advised that road works needs heavy machinery, like bulldozers, graders, loaders, rollers and trucks. Since you already have two tipper trucks you plan to buy one bulldozer D6, one grader G120 and one front-end loader. You are therefore getting quotations from second-hand dealers in the capital. The best offers you can get are:*

*bulldozer D6, 4 years old = 53 000 NU*
*grader G120, 5 years old = 47 000 NU*
*front-end loader, 6 years old = 43 000 NU*

*The total investment for the three pieces of equipment without spare parts, specialized tools and workshop equipment is 143 000 NU. The local bank offers you a loan for the purchase of this equipment at an annual interest rate of 18%, and repayment should take place within 5 years' time. A rough estimate shows you that you would have to pay the bank in the first year some 54 000 NU (143 000/5 yrs = 28 600, 18% of 143 000 = 25 740; 28 600 + 25 740 = 54 340). This is without having paid any operational costs, salaries and overheads for running the equipment. This seems to mean that the work may not be worth bidding for, as you have been informed that the new roads programme would give out contracts in the range of 100 000 to 150 000 NU per year and per contractor. Your uncle, who has many years of business experience, said 'Let's think again. Maybe we don't have to buy all this equipment. Is there not a better way?'*

As you know, the purchase of heavy equipment is expensive and in addition you have high costs of spares and support (due partly to the many different types of imported equipment). Furthermore, whereas spare parts can normally be obtained within 24 hours in industrialized countries, the process can take many months when they have to be imported to a developing country.

Then there is the question of availability (availability = the measure of time that a piece of equipment is available to do work compared to the normal working period; usually expressed as a percentage of the annual possible working hours). Specialist equipment can achieve high availability and performance in the hands of contractors in industrialized countries, and can repay its investment cost in a relatively short period of time. But in developing countries it is very difficult to achieve high availability.

*Example:*
*You have thought carefully about your uncle's advice, and have realized that you would run a big risk if you buy the equipment. First of all you are not even sure that you could win a contract; secondly you need a very good output from your equipment to be able to repay the loan and to cover all associated costs before you can make any profit; and thirdly you are not sure how good the equipment which you are being offered really is. So you decide to send your brother, the mechanic, to have a look at the equipment in the dealer's yard.*

*Your brother's investigations are not very encouraging. The bulldozer would require a new injector pump and the nozzles also need to be changed, which would cost you approximately 11 000 NU. The grader has hydraulic rams which are obviously leaking and the blade is completely worn out. A set of blades would cost you another 1500 NU and would need to be replaced every 2 to 3 weeks of full operation. The front-end loader seems to be in good order, although the hour-meter is not working and it is difficult to judge how the engine is performing.*

*To get the spare parts for the bulldozer and the grader would take at least 6 months from the time you buy the equipment. The agent for the spare parts told you that these spares need to be ordered abroad after he has obtained the necessary import licence as these spares are not very commonly used in the country.*

Good workshop and support arrangements are not usually available to local contractors in developing countries. Instead of achieving utilization rates in industrialized countries of 60 to 70% or more (1000 to 1200 hrs per annum), utilization in developing countries is typically of the order of 20% (350 hrs per annum) for heavy equipment like bulldozers and graders. (Utilization = the measure of time that a piece of equipment is actually used for work, compared to annual available working hours.)

*Example:*
*Now you are getting worried after realizing the hard facts. If you still go ahead and buy the equipment you would probably not be able to start operations for quite some time (which means paying bank interest with no income to set against it). Moreover, you fear that the equipment may break down at any time and more repairs would be necessary. This would destroy your reputation as well as your finances if you were given a contract and it overran. Think again about the wise words of your uncle: 'Is there not a better way?'*

## SOME BACKGROUND FACTS

In industrialized countries daily labour costs are relatively high (between 50 and 100 US$) and a contractor naturally tries to substitute equipment for labour whenever possible. In most developing countries the daily wage rate is much lower (between 1 and 5 US$), while there is also a high level of unemployment.

Table 2.1 shows some of the fundamental differences in resource costs and performance:

**Table 2.1** Typical resource features

| | *Industrialized countries* | *Developing countries* |
|---|---|---|
| Daily labour wage (US$) | 50–100 | 1–5 |
| Cost of an item of heavy equipment, e.g. a new motor grader (US$) | 100 000 | 150 000–200 000 |
| Annual utilization of motor grader (hours) | 1000 | 250–500 |

With low utilization, the heavy equipment will probably not complete its full useful life of 10 000 to 15 000 hours. It will probably become obsolete after about 12 to 15 years due to spares being no longer available. The high interest rates and frequent currency devaluations in developing countries mean that investment in heavy equipment is difficult to repay even if the equipment is put to good use.

You as a contractor will therefore have to think twice before you invest your money in specialized equipment that is relatively limited in its range of use.

*Example:*
*You read again the access and feeder road programme description which you were given by your contractor's association. It states that the programme managers intend to try to carry out the work by making maximum use of locally available resources, such as material and labour. This suggests that most of the work could be done using labourers instead of machines.*

*You know from your work as a building contractor that casual labourers cost you only 1.50 NU per day and that you can recruit them as required, provided you follow the national labour laws, of course. Moreover, there are job seekers outside your house every day asking for work.*

*From your experience as a building contractor you also know that using labourers is a very flexible approach to managing a project. If there is much work you can add a few labourers, and if the workload is reduced you can decrease the number of labourers. If a worker falls sick or leaves, you can simply hire another. Moreover, labourers are more versatile than equipment, and to provide a little training to an existing workforce is much cheaper than buying new plant and equipment.*

*Although using labour seems to offer an alternative, you are still doubtful whether heavy road works can be done by labour.*

*Therefore you want to investigate which of the road work activities can be satisfactorily carried out by labour and which must be done by equipment.*

# Technology Options

Road construction and maintenance technology in industrialized countries tends to be based on single activity, high cost, sophisticated items of equipment that maximize the use of their most expensive resource, labour. Alternative technologies make better use of the low cost and flexibility of the resource that is available in abundance in most developing countries – labour.

A number of developing countries have already shown that labour-based methods produce a high quality job for the client and profits for the contractor. Hand tools for the labourers can also be manufactured locally.

The quality of labour-based work can be comparable to equipment-based work, providing appropriate management systems are established. This is particularly important for setting out, geometric control, and quality control systems. Good performance and low costs require the establishment of suitable motivation and labour management systems.

Labour is often abundantly available close to the road where agriculture is established, and low or intermittent labour demand in that sector usually permits casual or more permanent employment on roads. The work can be broken down into simple activities which can mostly be carried out by unskilled labour.

With few other possibilities for employment in many rural areas, people will often walk daily up to 3km or more from their home to the site. It has been possible to recruit a workforce of 80 people in areas with a population density of less than 25 persons/km$^2$. The costs of labour – accommodation and transport – which are large components of equipment-based methods, are avoided.

*Example:*
*In the southern part of the country the Ministry of Works is carrying out a labour-based feeder roads project by direct labour (force account). You always thought that this is something the government does to provide employment in an area where many people cannot find work. As the government is doing the job, using their own organization, you have also assumed that productivity is not very high and there is no question of making a profit.*

*A friend of yours, who graduated from the polytechnic together with you, is employed by the Ministry of Works and you visit him to discuss the work, and especially your problems. After your discussions, and a field visit to the Ministry's project, you are surprised to see that the work is progressing well, and productivity is higher than you expected (although you can also see ways in which it could be improved further with good management). As a result you come up with the following list of observations on road construction using heavy equipment compared to labour-based methods.*

| *Construction/ maintenance activities* | *Heavy equipment method* | | *Labour-based method* | |
|---|---|---|---|---|
| | *Method* | *Application* | *Method* | *Application* |
| Setting out | Theodolite, surveyor's level | Good, also for difficult work | Boning rods, ranging rods, Abney level, line level | Good enough for simple setting out and control work |
| Bush clearing, tree felling | Bulldozer | Good, but seems to destroy the topsoil layer | Labour | Good, protective work possible |
| Vegetation and topsoil removal | Bulldozer, loader and trucks | Good | Labour with wheelbarrows/ animal carts/ tractors and trailers | Good |
| Boulder removal | Bulldozer | Good | Labour with crowbars and ropes | Good, but slow |
| Rock breaking | Drilling rig and explosives | Effective | Hand drills, plugs and feathers or 'Fire-water' method | Fair, slow |
| Excavation (soil or gravel) | Bulldozer, scraper, excavator | Good | Labour | Good |
| Loading (soil or gravel) | Loader, motor scraper, excavator | Good | Labour | Good, but only if loading height is less than 1.5m |
| Haulage | Motor scraper, trucks | Good | Tractor and trailer, animal-drawn carts, animal pannier, labour and wheelbarrows | Good, but maximum hauling distance for tractor 10km |
| Unloading | Motor scraper, tipper truck | Good | Labour or self-tipping device | Good |
| (Re)shaping | Motor grader | Good | Labour | Good, if soil is not too hard |

| Construction/ maintenance activities | Heavy equipment method | | Labour-based method | |
|---|---|---|---|---|
| | Method | Application | Method | Application |
| Compaction | Self-propelled (vibrating) roller | Good | Tractor towed deadweight roller, Hand operated vibrating roller | Good |
| Ditch cleaning and/or cutting | Grader | Good, gives a V-shaped ditch | Labour | Good, but requires a U-shaped ditch |
| Cleaning and minor repairs to culverts and bridges | n/a | Poor | Labour | Excellent |
| Building scour checks | n/a | Poor | Labour | Excellent |
| Grading unpaved surface | Grader | Good, but machine often under-utilized | Tractor-towed grader | Good, labour only cannot do well |
| Filling of pot-holes | n/a | Poor | Labour, bitumen handsprayer, pedestrian roller | Good |
| Grass cutting | Self-propelled or tractor-towed grass-cutter | Good | Labour | Good |
| Maintenance of road signs | n/a | Impractical | Labour | Good |
| Road line marking | Marking machine | Good | Labour | Fair |

*Analysing your list, you are now sure that there is actually a lot of potential for labour-based works, as many of the equipment activities can easily be substituted by labour.*

*Only the hauling of gravel and material cannot be done by labour in the project which you are interested in. Also the compaction would require some heavy equipment, but hand-operated rollers could do the job as well, and they would not be so expensive. They also have the advantage that you could use them on building sites to compact backfill, foundations, hardcore layers and so on. For the hauling of gravel and material you are thinking of using your own two trucks and if necessary you could hire additional trucks.*

*The financial commitments you would face when using a labour-based approach would be substantially less than buying the equipment you had planned for. All you would have to procure now are the two hand-operated rollers and enough hand tools for about 100 labourers. After some calculations you arrive at a total investment of about 12 000 NU. You could afford this without risking too much, and you would be able to repay such a loan within one year if you get one of the construction contracts. You also do not need to procure the tools too much in advance as they are available off the shelf at the local builders' merchant where you have an account. The two rollers will have to be ordered now, but you are ready to take this risk as the cost is not too high, and it will probably be possible to hire them to other contractors when you do not need them.*

*Knowing that you would have much less capital involved, which you would have had to borrow, the next question is – how much does the labour-based approach cost me and how does it compare to equipment operations? To answer this you analyse some typical activities and compare the two approaches with each other. One of the activities is 'excavation of soil to form the sub-base'.*

*First you need to know what a bulldozer (D6) would cost you per hour.*

*Assumptions*

| | | |
|---|---|---|
| *Expected economic life* | = | *8000 hrs over 10 years (assuming 40% utilization)* |
| *Annual usage* | = | *800 hrs* |
| *Current price new* | = | *146 000 NU* |

| *Cost Components:* | *Cost NU/hr* |
|---|---|
| *Depreciation, replacement, interest* | *= 19.90* |
| *Fuel* | *= 7.50* |
| *Oils and lubricants* | *= 0.50* |
| *Spares and consumable* | *= 9.50* |
| *Operator's wages* | *= 1.20* |
| *Overheads* | *= 5.90* |

*Total cost of operating Bulldozer (D6) = 44.50 NU/hr*

*The output of a Bulldozer (D6) is estimated to be 75m*$^3$*/hr and the cost for 1m*$^3$ *excavation is therefore about 0.60 NU.*

*Now you must calculate what it would cost you to excavate 1m*$^3$ *of soil by labour:*

*Assumptions*

*Labour costs per day = 1.50 NU*

| *Cost Components:* | *Cost NU/day* |
|---|---|
| *Labour wages* | *= 1.50* |
| *Hand tools (8%)* | *= 0.12* |
| *Supervision and Overheads (20% of 1.62)* | *= 0.32* |

*Total cost of Labour = 1.94 NU/day*

*The output of a labourer is estimated to be 4m³/day and the cost for 1m³ excavation is therefore about 0.50 NU*

*Labour-based technology begins to look promising – a difference in unit rate of 0.10 NU on rate of 0.50 NU would mean a 20% profit margin. You go on to calculate rates for other activities and obtain similar figures, so you become convinced that labour-based methods can not only be an advantage when it comes to saving capital interests but also when it comes to operational costs.*

## TRACTORS

At an intermediate level, agricultural tractors are the simplest, most robust, mobile mechanical power source available in developing countries for road work. They are much cheaper and simpler to maintain and more reliable than heavy equipment, and can be used to power a range of attachments for road sector activities. Tractor spares and repair-expertise are usually available in rural towns, whereas heavy equipment often needs to be taken to regional centres or the capital city for repairs.

*Example:*
*During your visit to the labour-based site of the Ministry of Works you have also realized that there is a lot of potential for the use of tractors. While lorries can only haul material, tractors can pull almost anything, such as trailers for material, bowsers for water or fuel, small graders for maintenance grading, drags for maintenance, rollers, and so on. Tractors, of course, are built for the use of farmers, and that also opens up the possibility of carrying out ploughing jobs on a contract basis. This flexible use of tractors fascinates you and you begin to regret that you earlier bought the two lorries, although they give you very good service, especially on long hauling distances.*

*As a dynamic contractor you are already making long-term plans and you are planning to invest money in tractors if the contracts will give you a good profit.*

As a small-scale contractor you will probably be attracted by maintenance operations, such as regravelling and routine maintenance. It is therefore important that you are fully aware of the general maintenance needs and work methods. Only then will you be able fully to meet the requirements of your clients. Road maintenance contracts, especially for routine maintenance, can be long-lasting and if you do the work well you have a good chance to get a steady amount of work over a period of time.

This is very helpful for a contractor because with a steady workload you can be more selective in bidding for projects, and will not have to 'buy' work to keep your workforce and equipment employed.

*Example:*
*After you have carried out your first road construction contract the client is very satisfied with your work and invites you to bid for a contract for routine maintenance activities on the road you have constructed. This provides a good chance to get a long-term contract. Although the profit margin is not very high, cash would flow in steadily every month. You could ask for a bank loan to buy a new tractor and trailer, knowing that the profit you get from your routine maintenance contract allows you to pay back interest and amortization.*

In many cases, costs of labour-based operations can be less than for those carried out using equipment, and for most work quality can be equal. The choice between equipment and labour-based methods affects the basic organization of road maintenance. Equipment-intensive work favours a more centralized organization, whereas labour-based work favours decentralization.

## Your Choice

Finally, you have to make your own choice as to which technology approach you want to make under the specific conditions in which you have to operate. Before you make your final decision it will be useful to answer a few basic questions:

- What is your financial position? Have you a substantial amount of money that you could invest?
- What are the conditions for bank loans? Would you be able to pay them back within a reasonable time and without stretching your other financial commitments?
- What are the prospects for civil works in the future? Could you realistically expect to obtain continuous contract work for equipment which you might use?
- What are the chances that such equipment would stand idle? Could you easily get spare parts and good service from the agents?

If your answer to all these questions is NO, then you should not choose to procure heavy equipment.

Here is another set of basic questions to ask yourself.

- Can you easily recruit labour in your area?
- Are the daily labour wage rates low?
- Can hand tools be bought locally?
- Are the costs per unit of output lower when using labour rather than equipment?
- Can you carry out most of the activities using labour instead of equipment?

If you can answer all these questions with YES then you can replace equipment with labour and you may only need some small equipment for those activities which you cannot do with labour, like hauling material and compaction.

# CHAPTER 3: SOIL MECHANICS

## Learning Objectives

After you have completed this chapter you should know:

- the principal soil classifications
- how useful the different soils are for road works, especially for gravelling
- how to interpret soil and gravel contract specifications
- how to carry out simple field tests in order to select the specified gravel
- the requirements for compaction of gravel surfaces.

## Introduction

On rural projects the soil will form the primary building material for roads. It is therefore important for you to know how to recognize which soils are suitable for road works and which are not.

Contract specifications often describe in specialist terminology the required quality for soils. As a contractor you have to understand these specifications in order to 'interpret' them in practical terms. You should also be capable of carrying out simple field tests which will allow you to make the required choice of suitable soils. This is particularly important for the selection of gravel. Although the client often indicates the source of gravel for a regravelling project, you will need to test the quality of the gravel before you actually use it.

In most routine maintenance and regravelling contracts it is the client's responsibility to provide a satisfactory source of gravel. However, it is the responsibility of the contractor to ensure that only good quality gravel is used in road surfacing. This means that you have to know enough about soil mechanics to be able to carry out your responsibilities under the contract.

Before your client designates the source of gravel, sample tests of the gravel quality are normally carried out. Even if these tests indicate a good quality gravel, there may be pockets of unsuitable soil in the quarry. You should be able to recognize these pockets. If the quality of the soil is not in accordance with required specifications, you should notify the client immediately. The client will then either have to give you a written authority to use the material with lower gravel standard or will have to make a variation in the contract so that you can use gravel from a different location. It is therefore important for a contractor to know how to determine the quality of gravel or soils; you could lose a lot of money if you make the wrong decision.

In this chapter you will learn about simple soil tests that you can carry out yourself to check on the suitability of various soils. As a ROMAR contractor you will not be involved in detailed laboratory tests. Those will be arranged by the client when detailed soil classification is required.

In addition to soil recognition and testing, you need to know how to compact the different layers of soil in order to achieve the required bearing-capacity of the road. This will be described in the contract specifications in terms of acceptable test results. You must be able to interpret these figures and take the necessary practical actions.

## How Soil is Described

To understand fully what the contract documents require in terms of gravel or soil specifications it is necessary for you to learn the general terminology employed. The following brief explanations are intended to give you a practical and easily understandable description of the terminology, which should be sufficient for your business purposes as a road maintenance and regravelling contractor.

### SOILS

*Gravel*
Strictly according to soil classification, gravel is defined as stones (2–60mm) but for road work use, a more useful definition is a mixture of stones (maximum 30mm), sand and clay.

*Sand*
A coarse-to-fine gritty soil with grains of size 0.06–2mm. Sand is normally firm when damp.

*Silt*
A soil with very small particles (0.002–0.06mm), which is powdery when dry but very soft when wet. For a quick test, when you roll a lump of silt between your hands they will not get stained.

*Clay*
This is a soil with even smaller particles (<0.002mm). It forms hard lumps when dry and the surface is cracked, but is sticky and soft when wet. For a quick test, your hands will be stained if you roll a lump of clay between your hands.

*Organic soil*
This is dull and dark in colour, and often has a distinct smell. Topsoil is almost always organic. Swamp soils usually contain remains of plants (fibres, roots, and so on).

## GRADING

*Well-graded*
Material with a wide range of particle sizes which are well distributed (Note: a mixture of particle sizes means that the soil will be easier to compact)

*Poorly-graded*
Material with too much of some sizes and too little of others

*Uniformly-graded*
Material with a limited range of sizes, mainly concentrated in one size category.

## OTHER CHARACTERISTICS

*Cohesive*
The particles of a soil stick together (mainly the clay fraction)

*Non-cohesive*
Does not stick together (mainly sand and gravel)

*Coarse-grain soil*
Mainly sand and gravel (little or no clay, little or no sand)

*Fine-grain soil*
Mainly silt and clay.

## SOIL CONDITIONS

*Density*
In a dense soil the particles are close together (or well-compacted)

*Compaction*
The process which packs the particles close together, and so increases the density

*Bearing capacity*
The strength of the soil (measured by the weight that can be loaded on to a specified area without penetration, or the amount of penetration under a certain load on a specified area)

*Plasticity*
Measures whether soil can be moulded and hold its new shape

*Permeability*
The degree to which water can penetrate a particular soil

*Optimum moisture content*
The water content that gives the best effect of soil compaction.

## The Surface Layer

A road is often built up of a number of different layers, such as the sub-base, base and surface layer, which spread the traffic load over the natural soil (subgrade). Roads which are built for the traffic of a few light vehicles, or on very good soils, might not need all these layers. Sometimes it is sufficient just to add a good surface layer on top of the natural soil. There are even cases where the natural soil is strong enough to carry the projected number of vehicles (*in situ* gravel).

Although they are relatively expensive, bituminous surface layers can be used to seal the surface and prevent surface water (rain) from penetrating and weakening the base and subgrade. In the case of a gravelled road this function is carried out partly by the gravel layer and partly by the camber of the carriageway.

The type (bitumen, gravel, stabilized soil or natural soil) and thickness of the surface layer should be determined by the expected traffic density and the type of the natural soil. However, other factors, especially the funds available and financial priorities, but also the location of the road and the availability of suitable material in the area, usually influence the design of the surface layer.

How suitable a soil is as a surface layer depends on how it behaves in dry, as well as in wet, weather. On the one hand, in dry weather a fairly high proportion of clay is desirable to bind the particles together and prevent corrugation. On the other hand, in wet weather the presence of a lot of clay in the mixture

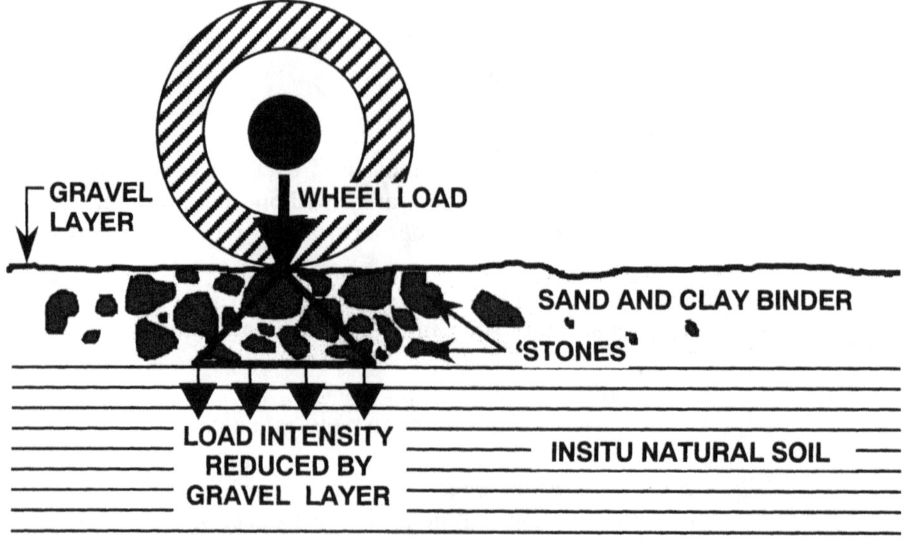


Figure 3.1 *Function of gravel surfacing*

makes the surface slippery, and ruts are easily formed. Therefore the specifications for soil used as surface layer has to be a compromise between the ideal requirements for wet and for dry weather.

As a rough guide, a mixture of approximately 50% stones (gravel), 40% sand and 10% clay will produce the best soil for a surface layer. The gravel will behave better if the stones in the mixture are of various sizes or 'well graded', so that the voids between the particles are filled. Angular-shaped particles are preferable, because they will 'lock' together better than round particles.

| | |
|---|---|
| | **larger stones in contact with each other** |
| | **smaller stones in spaces between larger stones** |
| | **sand and clay as binder (glue) in between** |

Figure 3.2 *Well-graded gravel layer*

In general, a slightly higher clay content is required in a dry area to compensate for dust loss. In wet areas a high clay content makes the road slippery during rain. Pure clay or silt, mixtures of silt and clay, and organic soils are generally unsuitable for surface layers.

The stony particles to be used in a surface layer should not be bigger than 30mm in order to obtain a reasonably smooth finish. Larger particles should be crushed by hand or by roller.

## Soil Identification Procedure

This Section provides you with an introduction to the Unified Soil Classification System (USCS) which is used when classifying soils in most countries. You do not need to be an expert in the entire procedure, but you should know enough to see how the results of your simple field tests measure the suitability of various soils that you may come across.

The detailed tests for gravel selection are usually carried out by the client in a laboratory. Although more scientifically correct and detailed, they follow the same principle procedures that are outlined in Table 3.1. The path for clayey gravel (GC) is highlighted for easy reference (this is the gravel which is most suitable for surfacing roads). Remember always to consult this table when comparing your technique with the description of the field tests methodology described below. When undertaking these tests, you follow a step-by-step approach that gradually makes it possible for you to check that all necessary criteria for suitable gravel are fulfilled.

## Simple Field Tests

These simple field tests will allow you to determine the quality of gravel in a sufficiently accurate way if laboratory tests cannot be carried out. We will present it step by step so that it is easier for you to follow; you will also be given a chance to test your skills at carrying out a test like this in the Workbook. These tests follow the highlighted procedure to select gravel as explained in Table 3.1. It is important that you relate your test results to this table before you move to the next step.

### STEP 1

*Settling Test (or Bottle Test)* – for illustrations see page 33
After you have taken a sample from the gravel, according to Table 3.1 you first have to determine whether your sample has a coarse grain content of over 50%. Coarse grains are stones and

## TABLE 3.1- SOIL IDENTIFICATION PROCEDURE

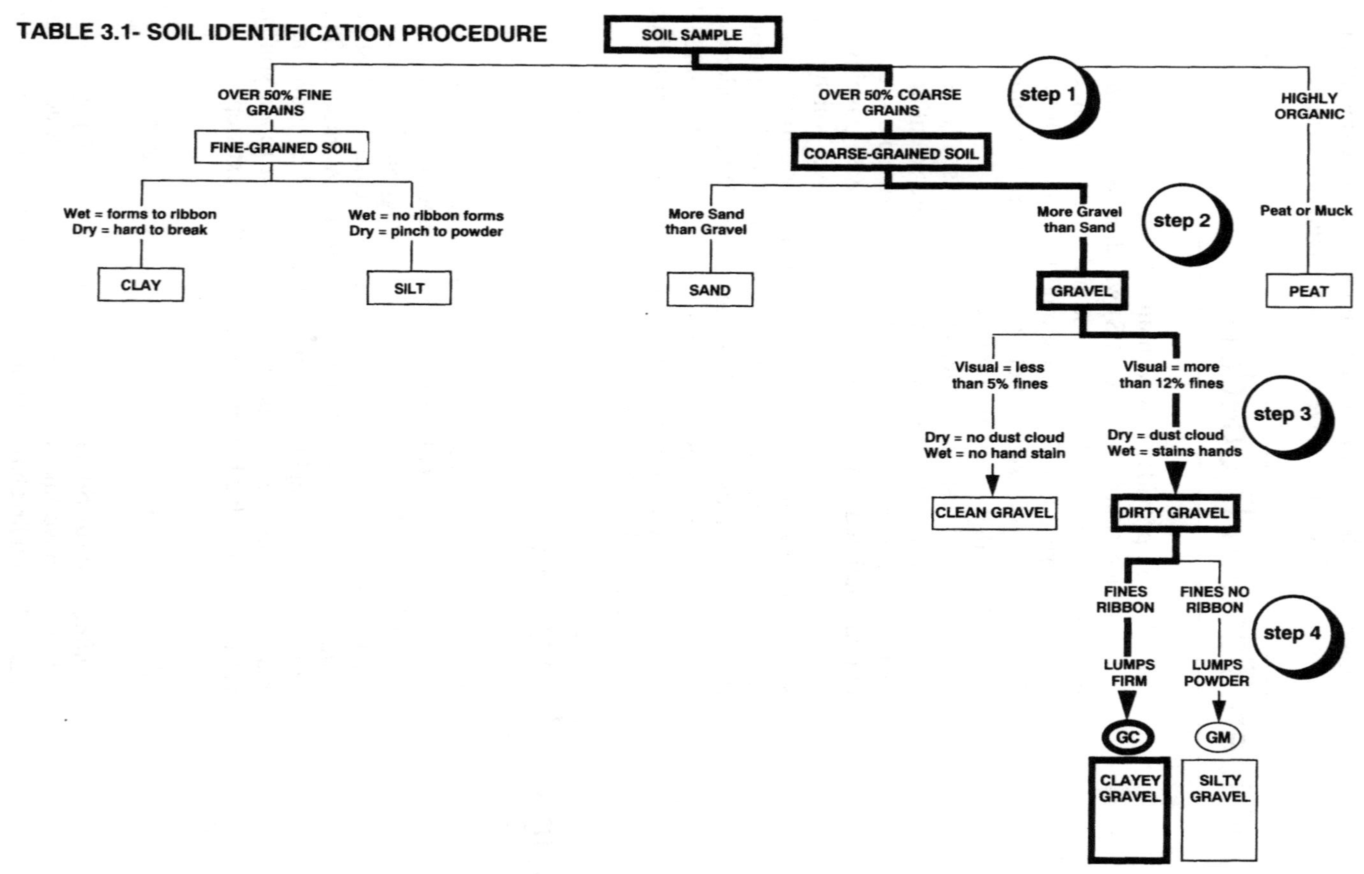

*Procedure:* (see step 1)

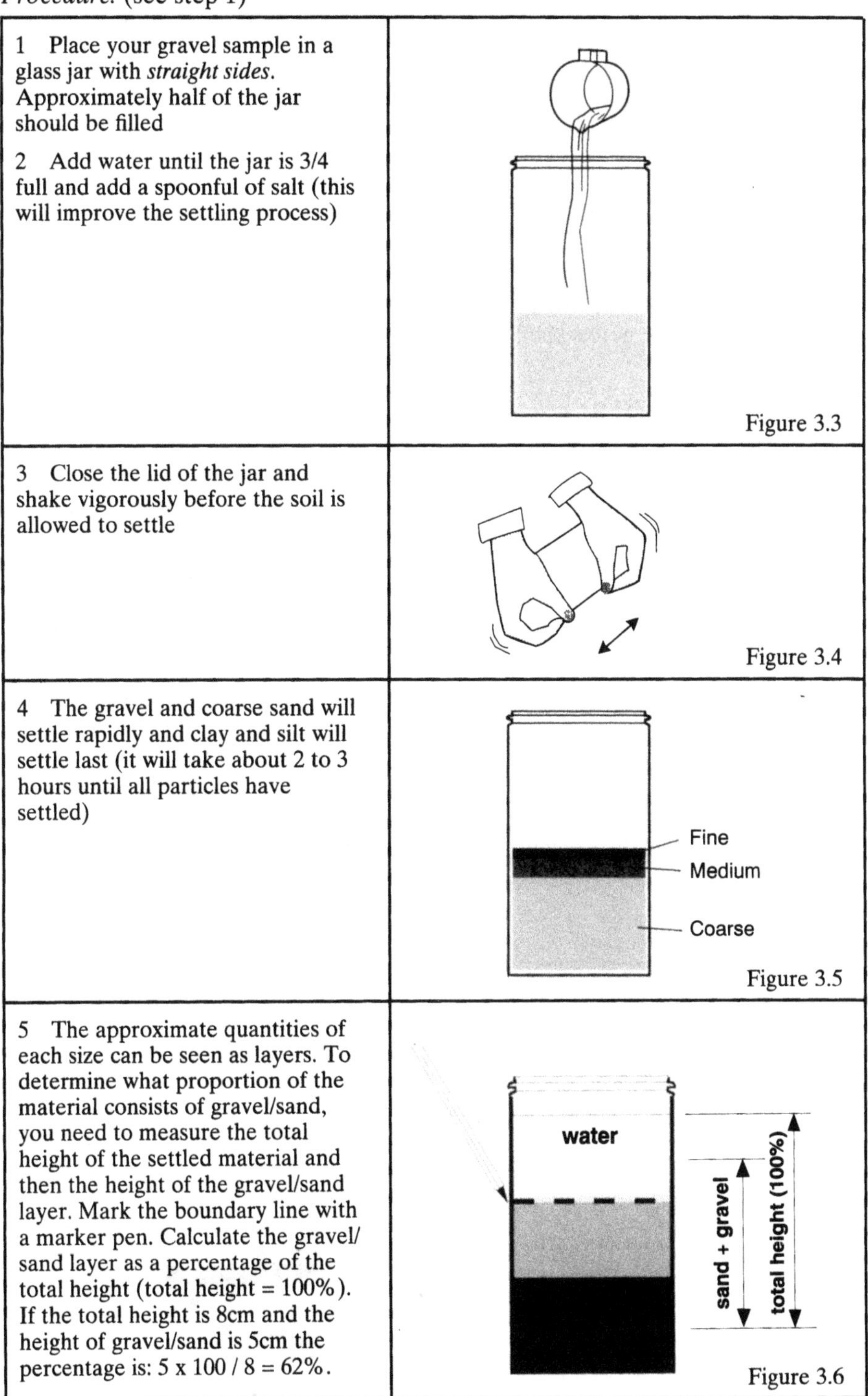

1 Place your gravel sample in a glass jar with *straight sides*. Approximately half of the jar should be filled

2 Add water until the jar is 3/4 full and add a spoonful of salt (this will improve the settling process)

Figure 3.3

3 Close the lid of the jar and shake vigorously before the soil is allowed to settle

Figure 3.4

4 The gravel and coarse sand will settle rapidly and clay and silt will settle last (it will take about 2 to 3 hours until all particles have settled)



Figure 3.5

5 The approximate quantities of each size can be seen as layers. To determine what proportion of the material consists of gravel/sand, you need to measure the total height of the settled material and then the height of the gravel/sand layer. Mark the boundary line with a marker pen. Calculate the gravel/sand layer as a percentage of the total height (total height = 100%). If the total height is 8cm and the height of gravel/sand is 5cm the percentage is: 5 x 100 / 8 = 62%.



Figure 3.6

| Decision: | If the percentage of coarse grains is over the 50% limit you know you have coarse grained soil → continue with tests<br><br>if the percentage of coarse grains is under the 50% limit you know you have fine-graded soil → you can stop testing now; the sample shows that the soil is not gravel |
|---|---|

sand with a size of 0.02mm and above. Without using sieves it will be difficult to identify exactly the borderline between silt/ clay and sand/gravel. However, the bottle test will allow you to find out whether the coarse grain content is clearly above or well below the 50% limit.

## STEP 2

*Sand or Gravel?*
The second step is to find out whether your sample consists mainly of sand or gravel.

You will remember that the best soil for a surface gravel layer is 50% stones (gravel), 40% sand and 10% clay. Therefore you need to find out where the boundary line between sand (0.06–2mm) and gravel (2–30mm) is in your test jar. After you have calculated the percentage of the sand and gravel layers in relation to their total height you will then be in a position to say whether your sample is mostly sand or mostly gravel:
If the percentage of sand is higher than gravel, the sample can be described as sand.

If the percentage of gravel is higher than sand, the sample can be described as gravel.

*Procedure:*

| 1 Define the borderlines between gravel and sand, and between sand and fines on your sample jar. Mark them with a marker pen | 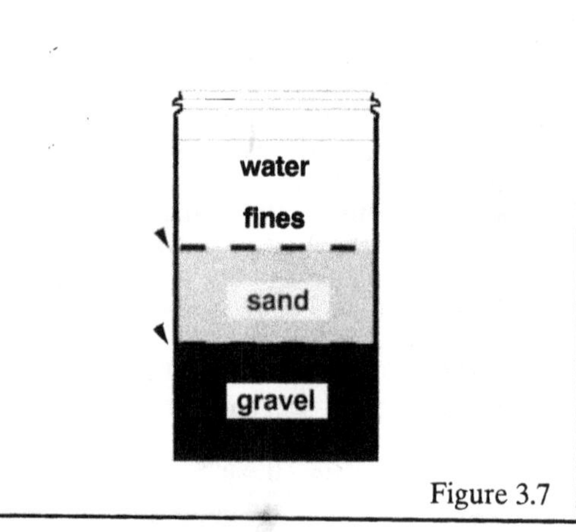<br><br>Figure 3.7 |
|---|---|

| | |
|---|---|
| 2 Calculate the percentage of each, the gravel and sand layers, in relation to the total height of the sample in the jar<br>Gravel % = (G/T) × 100%<br>Sand % = (S/T) × 100% | 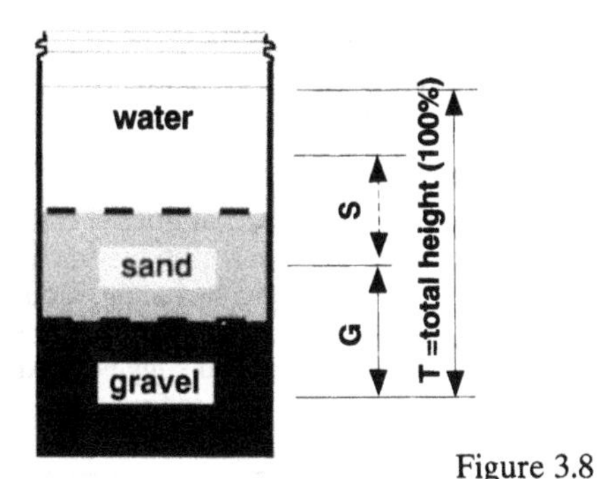<br><br>Figure 3.8 |
| Decision: If the percentage of gravel is clearly over the 50% limit you know you have gravel → continue with tests<br>if the percentage of gravel is clearly below the 50% limit you know you have sand → you can stop testing now; the sample shows that the soil is not gravel | |

## STEP 3

*Clean Gravel or Dirty Gravel?*

Following the path highlighted in Table 3.1 we know that our ideal gravel needs to fall under the category of 'dirty gravel' (a 'clean gravel' with no clay or silt is ideal for making concrete, but for road building, you need the fine material to provide the right mixture). In this step we need to find out the actual percentage of the fines in relation to the total sample.

Knowing the percentage of the sand/gravel layer as calculated in step 1, it is now easy to calculate the remaining percentage for the fines. If the fines are well above 10%, we know that we have dirty gravel.

*Procedure:*

| | |
|---|---|
| 1 Calculate the percentage remaining for fines by deducting the percentage of the gravel/sand layer, as identified in step 1, from 100% | 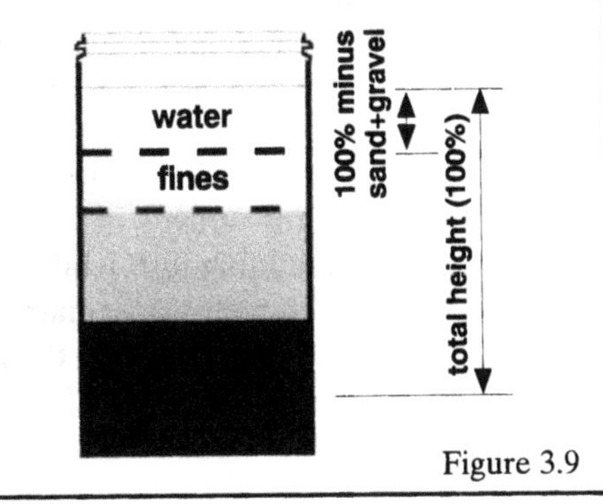<br><br>Figure 3.9 |

Decision: If the percentage of fines is over 12%, you know you have dirty gravel → continue with tests.

Remember: If the percentage of fines is *much* over the 12% limit then you know that your sample has too much clay or silt and you should ask your client for advice.

If the percentage of fines is clearly below 10% you know you have clean gravel → you can stop testing now; the sample shows that the gravel is not suitable for road surfacing. You need to inform the client. If there is no better gravel available in the area you may ask for permission to use this gravel or you could suggest mixing with clayey material from the roadside. Whatever solution is found, ask the client to confirm it in writing and make sure you are paid for any additional work required.

## STEP 4

*Clayey Gravel or Silty Gravel?*

If your analysis has been successful you can stop here, knowing that these simple tests have allowed you to select a gravel which is suitable for road surfacing.

However, it is sometimes necessary to find out whether the material is a Clayey Gravel or Silty Gravel (see Table 3.1). The clay in the gravel binds the gravel particles together and ensures that the gravel layer remains firm. Silt becomes powdery when dry and does not allow the gravel particles to stick together. In order to discover this you need to test the fines in a separate exercise. You therefore need a sample of the fines only. In most cases you will not have a standard sieve to separate the fines from the coarse material. The easiest way of getting the fines is to keep the sample jar in the sun until the water has almost evaporated. With a spoon you can then carefully remove the fines from the top. With this moistened material you can then carry out two different tests to find out whether it is mostly clay or mostly silt. Both tests check what is called the plasticity of the material. If the material has a high plasticity, we know the material is clay which can expand and shrink without disintegrating into powder. This material can bind the gravel particles together even when it is dry. If the plasticity is low, the material is silt. Silt cannot bind the gravel particles together so the bearing capacity of the gravel is reduced. One of the tests is known as the 'moulding test' while the other is usually called the 'drying test'. Both tests are explained in the procedure below.

Remember that the results of all these tests provide you with good indications of a soil's suitability. However, laboratory tests are always necessary when detailed soil classification is required.

*Procedure:*

| | |
|---|---|
| 1 Leave the sample jar in the sun until almost all the water has evaporated but the material is still slightly moist. Remove the fines carefully from the top of the sample with a spoon. | Figure 3.10 |
| 2A Moulding Test:<br>Try to form ribbons of threads with your hand on a smooth table or board. The material must still be moist for you to be able to do this.<br>If it is clay you will be able to form ribbons and the soil will stain your hands.<br>If it is silt the ribbons will crumble, or will be very short, and the soil will not stain your hands. | <br>Figure 3.11 |
| 2B Drying Test:<br>Make sure that the fines are still moist. Fill the material evenly into a matchbox and let it dry.<br>If it is clay it will crack and shrink.<br>If it is silt it will not crack and shrink but will tend to crumble when touched | <br>Figure 3.12 |

# Compaction

Your contract documents require that you should compact the surface layer of a road to:

- reduce hollow spaces
- increase bearing density
- reduce settlement
- reduce permeability
- increase shearing strength.

Compaction results in a dense surface which can carry substantial loads without getting depressions. It also helps to ensure that rainwater cannot easily penetrate the layer and soften the gravel or the base.

Compaction, in simple terms, packs soil particles closer together, resulting in a more dense and stabile soil. A volume of soil is composed of three components:

- solid soil particles
- water
- air.

Air does not contribute to the strength of the soil, and reduces its stability by allowing water to 'circulate' between the particles.

A certain optimum quantity of water can be useful, because it makes the soil easier to compact and contributes to the soil's strength and stability (optimum moisture content is different for each soil type and usually varies between 8% and 10% of the total volume). Water lubricates the particles and allows them to settle in a dense mass. Too much or too little water is bad. Less water is required for well-graded gravel/sand/clay mixtures. Where the natural (existing) moisture content is well below the required optimum moisture content, water must be added before effective compaction can be carried out. Where the natural moisture content (soaked soil) is significantly above the optimum moisture content the soil should be left to dry until the optimum is achieved and compaction can take place.

Therefore you, as a gravelling contractor, need to have a water bowser to be able to add water to the gravel when required.

It is also important for you to check the contract specifications carefully about the required bearing-capacity you have to achieve. It will be your goal as a contractor to find the most economical way to achieve the required standards.

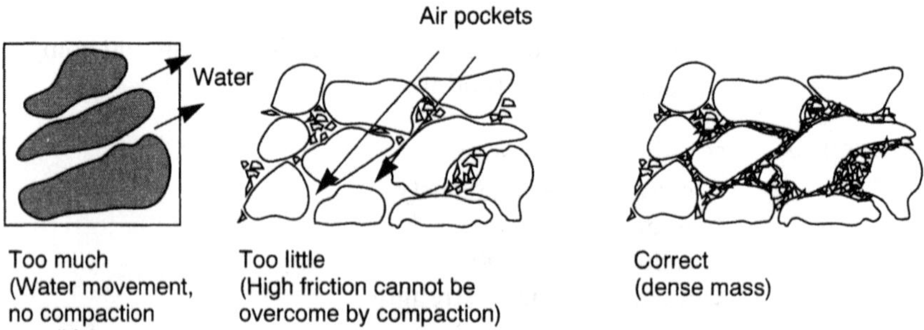


Figure 3.13 *Compaction*

For a small-scale contractor working on labour-based regravelling projects there are three principal ways of compacting gravel layers:

- compress the soil with a heavy roller which can be pulled by hand or towed by a tractor (non-vibrating rollers)
- use manually or mechanically operated tampers or rammers (they are more useful for small works, like culvert backfill)
- compress the soil with a vibrating roller which has a more powerful effect in compressing soil due to mechanical vibration (pedestrian operated vibrating rollers are sufficient for earth and gravel roads).

The required state of compaction is normally specified relative to a laboratory compaction test. For example, compaction to 95% means that the dry density of samples taken in the field should be 95% of the dry density obtained in specified laboratory compaction test.

Additional information about compaction for gravel surface layers can be found in Chapter 8.

# CHAPTER 4 EQUIPMENT AND TOOLS

## Learning Objectives

After you have completed this chapter you should know:

- the most appropriate equipment and vehicles to be used for routine maintenance and regravelling work
- how to plan and organize the service and maintenance of equipment and vehicles
- how to plan and control the use of equipment and vehicles for regravelling and routine maintenance work
- the required specifications for suitable hand tools
- what criteria to use when selecting hand tools and how to maintain them.

## Introduction

Construction plant is costly, so you will prefer to use labour whenever possible. However, some tasks require equipment and vehicles either to support labour activities or where labour-only work cannot meet quality standards. In general, equipment is necessary for hauling gravel and material and vehicles are required for efficient work supervision. The selection and utilization of equipment is a very important aspect of construction management, with the objective of obtaining a good return in terms of productivity and profit.

Labourers are your main means of production and we shall therefore dedicate Chapter 10 to discussing management of people. In order to enable your labourers to work effectively you need to provide them with adequate hand tools. This aspect is often underrated by project managers and contractors not fully appreciating how important is the quality of the hand tools, both for productivity and safety. Remember that each job requires a correct set of tools. As you know, it is very difficult to repair a wrist-watch with a sledge-hammer.

For road works, it has been argued that normal agricultural hand tools will be sufficient. The fact is that agricultural hand tools are normally far too weak and cannot withstand the tough conditions of a road site. The specifications for handtools must therefore be the same as for the rest of the construction industry. Every labourer should have the right tools of the correct quality for a particular task, and it is your job as contractor to make sure that they get them!

# Plant for Routine Maintenance and Regravelling

## REQUIREMENTS FOR ROUTINE MAINTENANCE

Routine maintenance activities on roads with low traffic densities, where grading of the carriageway is not necessary, are usually fully labour-based and require no support from equipment. The plant you need is limited to vehicles for the transport of tools and materials, and for the supervision of the labourers and inspection of the roads.

For routine maintenance contracts of less than 20km of roads where you would not usually have more than 10 labourers, no pick-up or motorcycle is normally required. The amount of supervision work on a stretch like this cannot justify such an investment.

Up to 20km supervision can easily be carried out with a bicycle.

It may not be necessary for the contractor to provide the bicycles, and gangleaders themselves could provide them against an adequate allowance. This has the advantage that the contractor is less bothered with repair and maintenance of the bicycles.

Transport of materials and tools can be done with hired transport. For example a tractor and trailer could be hired for a day from a local farmer to transport sand, stones and cement to a structure which requires repair, or to distribute new hand tools to the labourers.

Figure 4.1 *Vehicles for routine maintenance contracts (up to 20km)*

The following is a proposed list of vehicles for a routine maintenance contractor who looks after 20km to 100km of roads with approximately 10 to 50 labourers:

One pick-up for the transport of hand tools and small materials, such as cement bags or culvert rings. The vehicle can also be used for inspection tours.

The pick-up will not be required full-time and can therefore also be used for other work.

If you have no other work on hand in the area, it might be cheaper to hire transport instead.

One motorcycle 125cc to 175cc for the daily site supervision of the labourers by the road foreman. The motorcycle should be fitted with a carrier to allow the foreman to carry some measuring aids (template, strings, spirit-level and so on). The foreman should also be provided with protective clothing, crash-helmet, goggles and gloves.

For routine maintenance contracts larger than 100km an extra motorcycle will usually be required.

Bicycles will be necessary for the gangleaders (each covering approximately 20km) to carry out the daily supervision effectively.

Figure 4.2 *Vehicles for routine maintenance contracts (20–100km)*

For routine grading of the carriageway no motor-grader is needed as this work can be carried out with a tractor-towed grader or, on sandy soils, even with a drag. A tractor-towed grader has the advantage of flexibility. The tractor can be used for work other than grading, such as towing trailers, rollers or water bowsers. However, the investment for such a set may only pay off for larger contracts, such as routine grading of more than 100km of roads. Light towed graders (2 tons) can be operated with a 60hp tractor while heavy towed graders (up to 5 tons) require a tractor with 4-wheel drive and 100hp.

Figure 4.3 *Vehicles for routine maintenance contracts (grading)*

For routine maintenance activities on roads with high traffic densities, where grading of the carriageway is necessary, additional equipment will be required.

## REQUIREMENTS FOR REGRAVELLING

Regravelling work requires several pieces of equipment, which means a heavy investment for a contractor. If available, hiring or leasing can be an alternative, especially when it is unlikely that continuous contracts of this kind will follow. As a contractor you will have to carefully assess the likeliness of further gravelling work before you start buying equipment.

From an operational point of view the choice of hauling equipment depends mainly on the hauling distance. For hauling distances up to 8 or 10km, tractor–trailer combinations are usually the most economic alternative, while for distances of more than 10km, lorries and tippers could be a better choice, with their significantly higher transport capacity.

However, as you know, a contractor must first use what he or she already has available. For example, you are asked to tender for a regravelling contract with an average hauling distance of 5km. Among your equipment you have one flat-bed lorry and two tippers that are used only one day per week on another contract. Although it might not be the ideal hauling equipment for this distance, it will probably be more economic to use your own idle equipment than to buy or hire new tractors and trailers. Later on, if more jobs of this kind come in regularly, you might be in a position to acquire some tractors and trailers.

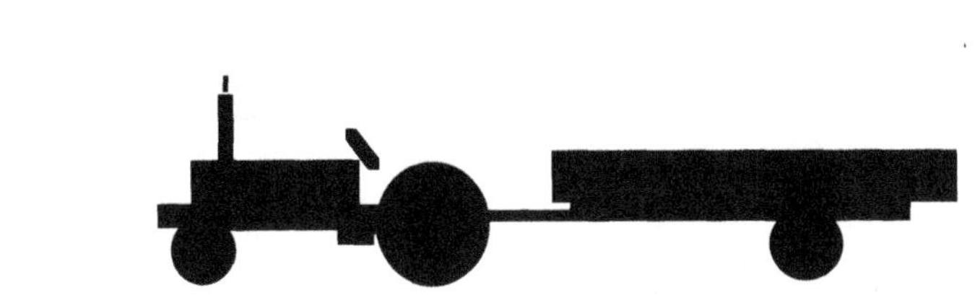

Tractor and trailer combinations are usually best suited for hauling distances up to 10km. Each tractor should have two trailers so that the tractor can be optimally utilized, e.g. one trailer is loaded in the quarry while the other is hauled to site. For an average regravelling site (e.g. 10km regravelling) 3 tractors and 6 trailers will be sufficient to complete the work in a reasonable time. Tractors for this sort of operation should be in the power range of 50 to 60hp while trailers should have a loading capacity of 3m$^3$.

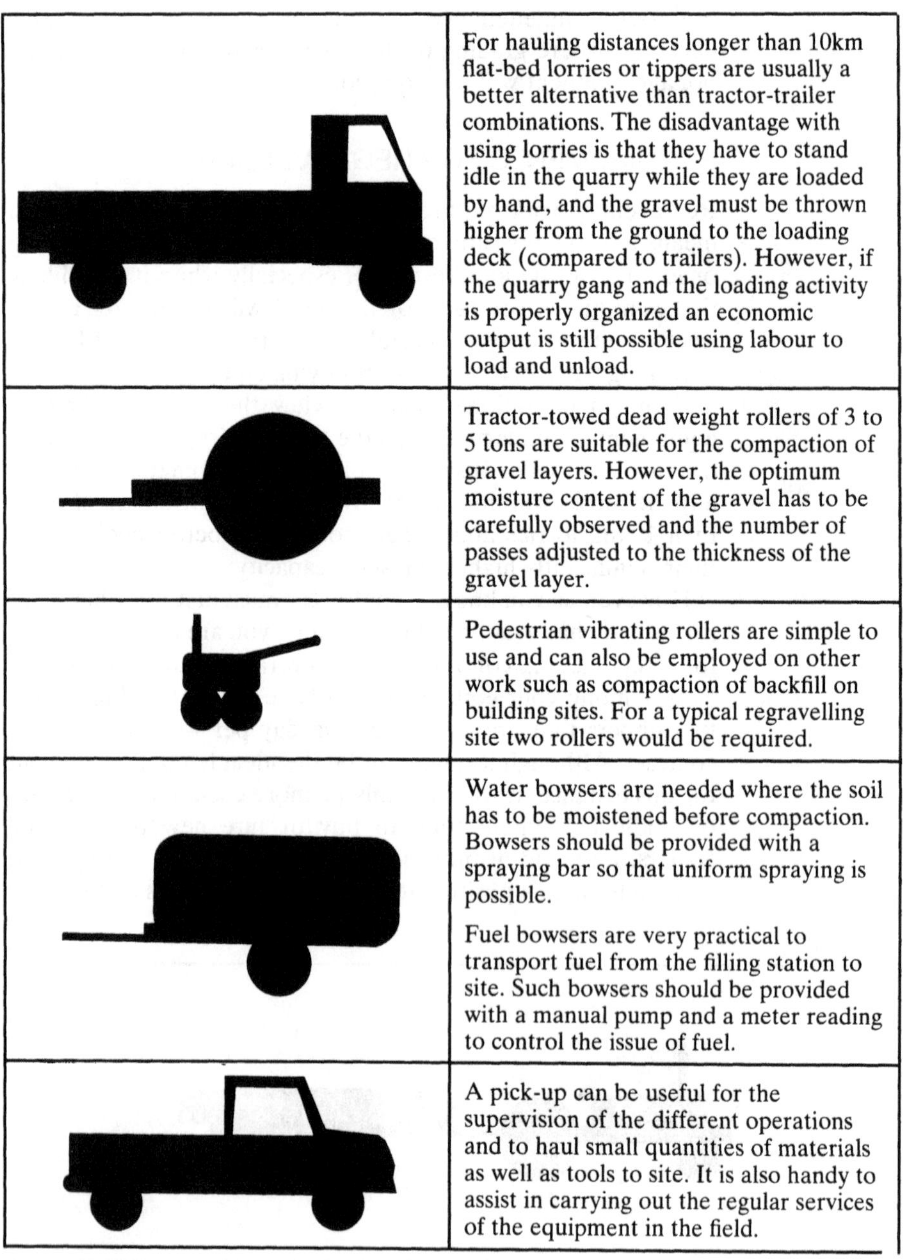

| | |
|---|---|
| | For hauling distances longer than 10km flat-bed lorries or tippers are usually a better alternative than tractor-trailer combinations. The disadvantage with using lorries is that they have to stand idle in the quarry while they are loaded by hand, and the gravel must be thrown higher from the ground to the loading deck (compared to trailers). However, if the quarry gang and the loading activity is properly organized an economic output is still possible using labour to load and unload. |
| | Tractor-towed dead weight rollers of 3 to 5 tons are suitable for the compaction of gravel layers. However, the optimum moisture content of the gravel has to be carefully observed and the number of passes adjusted to the thickness of the gravel layer. |
| | Pedestrian vibrating rollers are simple to use and can also be employed on other work such as compaction of backfill on building sites. For a typical regravelling site two rollers would be required. |
| | Water bowsers are needed where the soil has to be moistened before compaction. Bowsers should be provided with a spraying bar so that uniform spraying is possible.<br>Fuel bowsers are very practical to transport fuel from the filling station to site. Such bowsers should be provided with a manual pump and a meter reading to control the issue of fuel. |
| | A pick-up can be useful for the supervision of the different operations and to haul small quantities of materials as well as tools to site. It is also handy to assist in carrying out the regular services of the equipment in the field. |

Figure 4.4 *Equipment for regravelling work*

A practical approach to these decisions is necessary. You should always look for equipment which can be used on a wide variety of projects. Plant is only economic when it is being employed to earn money, so full utilization of equipment must be the overall goal of a contractor.

Figure 4.4 shows a range of possible alternative hauling and support equipment which can be used for regravelling work when employing labour-based methods. It will be up to you to choose the set which suits your business best and it is therefore necessary for you to carry out a careful financial assessment of the different alternatives.

# Maintenance and Repair of Equipment and Vehicles

## MAINTENANCE AND REPAIR ARRANGEMENTS

Everything you do as a contractor should either make you money or save you money. Plant maintenance costs money immediately, but saves you much more in the long run. Let us look at an example illustrating some of the problems.

*The New Construction Company (NCC), a small-scale contractor, entered the labour-based road regravelling business three years ago. After one year, when George Johnson, owner of NCC, had made a thorough assessment of the possible future market he decided to buy a new lorry. When George bought the lorry, the garage stressed that the manufacturer recommends that engine oil and filters are changed every 10 000km if the lorry is used on normal roads. However, if it is used on dusty roads, the service should be undertaken every 5000km. NCC is using the lorry for gravelling work where a lot of dust is created.*

*The recommended service costs 175 NU each time. George thinks that 175 NU is very expensive and besides, he is convinced that a service every 10 000km is sufficient. After two years and 58 000km the engine of the lorry requires a complete overhaul. This repair costs 4500 NU. Usually a lorry engine, if properly maintained and serviced, would not require such an overhaul at all, although a minor overhaul, like changing the piston rings and engine seals, would be required after 100 000km.*

*Although the cost of services was reduced by 1050 NU [at 58 000km, service every 5000km would have meant a total cost of 1925 NU (11 x 175) compared to 875 NU (5 × 175) with service every 10 000km; 1925 – 875 = 1050], the cost of the unnecessary repair was much higher. On top of these direct costs there is also a loss of time because the lorry was standing in the workshop for the overhaul, when it should have been used at a site.*

When it comes to machinery, your main interest as a contractor is, of course, to get as much output as possible from your investment. This thinking is not wrong as long as you appreciate the necessity for proper maintenance. It must therefore be your aim to set up a service and repair system which is both effective and cheap.

In setting up your system you have to ask yourself some important questions:

- What type of machinery and how many of each type do I have to maintain?
- What kind of facilities, mechanics, equipment and tools are needed to carry out services and repairs, and what do I have access to myself?
- How much would it cost to carry out the services and repairs myself and how much would it cost to have it done by an agent or garage?
- If I cannot carry out services and repairs, is there an agent or garage nearby who could do the work?

There is never an easy answer to these questions. However, here are a few tips to help you find the right solution for your company:

- The more plant you have, the more economic it will be to set up your own repair workshop. If you just have a pick-up and one or two motorcycles it will be cheaper to go to a garage. However, if you have enough equipment, it is worth considering carrying out regular services and minor repairs in your own workshop. The investment in facilities and tools need not be too heavy, and you gain extra flexibility by having direct control of the facilities. Guidelines on facilities, required resources and systems are given in the section Maintenance Organization, below.
- If you intend to set up your own equipment and vehicle service and repair facilities you have to consider very carefully the financial implications of such an undertaking:
  - what exactly do you need to procure; buildings, tools, workshop equipment? (Prepare a comprehensive list, and obtain cost estimates of all items.)
  - have you any capital at hand to invest?
  - if you have to borrow money, will you be able to pay the interest and repay the loan?
- Before you establish your own workshop you also need to estimate the potential workload for your company in the future. Will you have continuous work where you can utilize your equipment and vehicles, and consequently also your

workshop? If this is not the case then it is probably not cost-effective to invest money in a workshop. On the other hand, you may also be able to carry out service and repair work for other companies or persons who own equipment and vehicles. Whatever the situation, it will be necessary to carry out a careful survey before you make your decision.

- Even if you have your own service and repair facilities, it is usually cheaper and less risky to get specialists to carry out major repairs. In most cases you would not have the right tools and workshop equipment to undertake major repairs anyway, so the work should be given to a specialized agent or garage. If there are several companies which could carry out the job, you have to obtain quotations from them. Always ask for a cost estimate before you have a repair carried out.
- Whether you maintain your machinery yourself or use an outside garage you need to develop a detailed service schedule for your equipment and you must make sure that this schedule is strictly followed (see Service Schedule, page 53).
- Whatever system you choose, remember that regular routine maintenance saves costly emergency repairs.

## MAINTENANCE ORGANIZATION

If you decide to set up your own service and repair unit, there are some basic rules you need to remember:

- try to have only one make of each type of equipment, For example, all tractors should be of the same make. Standardization enables you to:
  - procure spare parts more economically
  - stock essential spares in bulk
  - have one set of special tools only
  - train your mechanics better by concentrating on one make.
- define a clear workshop organization and plant control system, including:
  - allocating one driver or operator to each piece of equipment and make him or her fully responsible for it
  - establishing daily checks and regular maintenance to be carried out by drivers or operators
  - defining clearly the type of repair work your mechanics are supposed to carry out and what should be given to an agent or garage
  - defining the activities that should be carried out on site so that the equipment does not need to waste time by travelling from site to the workshop

- defining the staff, tools and spares required for the workshop (see following section)
- establishing a reporting system that allows you to analyse carefully the workshop activities and mechanical services in the field (see later section)
- defining a detailed service schedule for all your equipment and informing your site supervisors, drivers, operators and mechanics about it (see later section).

## WORKSHOP FACILITIES, STAFFING AND ACTIVITIES

A simple workshop with essential equipment and tools will probably be necessary on larger regravelling contracts. The requirements for such a workshop should be:

- a roof covered area of about 50 to $60m^2$
- one inspection pit
- cemented workshop floor
- a small store for spare parts, tools and equipment

The level of the workshop floor should be kept above the surrounding ground, which must be well drained. If possible, electricity and water should be provided. The store must have lockable doors. To provide sufficient light in the workshop some transparent roofing sheets can be installed. Such a workshop would be sufficient to maintain all equipment that is required for a regravelling contract.

**Table 4.1** Work at base camp workshop

| *Suggested work to be carried out in your workshop:* |
|---|
| regular services |
| suspension repairs |
| steering system repairs |
| electrical system repairs |
| fuel system repairs, except fuel pumps |
| brake system repairs |
| removal and replacement of exchange units |
| engine tuning |
| tyres and tubes repairs and exchange |
| other minor repairs |

To be able to carry out these activities you would need to employ one qualified mechanic and probably one or two assistants.

It is also suggested that the regular service for the field equipment is carried out in the field, where the distance from site to your camp is more than 50km. Therefore you would need to equip your mechanic with transportable tools and equipment. Your pick-up can be used to transport the tools, spares and mechanics to and from the site.

**Table 4.2** Work at site

| *Suggested work to be carried out at site:* |
|---|
| regular services if distance to main camp is too far |
| daily inspection, cleaning and greasing |
| tyre and tube repairs |
| wheel-bearing adjustment |
| battery checking and filling |
| replacement and tightening of nuts and bolts |

Daily inspection, greasing and cleaning should be done by the drivers and mechanics. If you only have a few pieces of equipment on site, it is also advisable that you train your drivers and operators to carry out simple repair work on their own. If they are capable you could consider paying them some overtime, especially if that saves you from having to employ a mechanic. However, on a larger site it will probably pay to have a reliable mechanical assistant employed.

The following tables set out the suggested workshop and field equipment that is necessary to carry out the work at your base camp workshop and on site. These lists may not be complete, or they may contain items that you do not need for the equipment

**Table 4.3** Workshop equipment and tools

| | |
|---|---|
| 1 | portable arc welding kit with a 220V AC power outlet. For the welder – an apron, gloves, safety goggles, etc. |
| 1 | gas welding and cutting outfit |
| 1 | anvil – forge, if tools from the field, like shovels, picks etc. are to be repaired at the district workshop. |
| 1 | air compressor (not absolutely necessary) or foot pump |
| 1 | angle grinder |
| 1 | hardwood work-bench |
| 1 | hand drilling machine with 16mm chuck and variable speeds |
| 1 | torque wrench 50 – 225Nm |
| 1 | circlip pliers external 25 – 90mm diameter |
| 1 | circlip pliers internal 25 – 90mm diameter |
| 1 | drill set 2–26mm diameter |
| 1 | set of taps according to the fleet (metric/imperial/UNF/BSF) |
| 1 | pop rivet set |
| 1 | set of bearing pullers |
| 1 | multimetre = Volts/Amps/Ohms for DC up to 36/6000/10 000 |
| 1 | battery service and testing kit |
| 1 | battery charger |
| 1 | soldering iron |
| | common mechanical tools and tool boxes |

**Table 4.4** Field equipment and tools

| |
|---|
| complete spare tyres for all equipment on site |
| common mechanical tool-box |
| tyre levers |
| foot pump |
| axle stands |
| hydraulic jack 5 tons |
| strong ropes |
| winch (hand-operated) and steel ropes |

you own, but they should assist you in establishing your own workshop. However, if you set up your own workshop it will be necessary for you to seek advice from an experienced mechanic. To get high quality service from your mechanics you must let them have tools that fit your equipment. Trying to save money by buying second-class tools is false economy, but remember, if you have a standardized equipment fleet you need to spend a lot less on procurement of tools compared to having to service several different makes.

Your spare-part requirements, of course, depend on what kind of repair and service you undertake, and the type of vehicles and equipment you own. You will have to decide what spare parts to keep on the basis of your own assessment of your plant holdings and your own experience, remembering that fast-moving items required for routine services must be available at all times.

The most important items to be kept for all vehicles and equipment at the site workshop are listed in Table 4.5.

**Table 4.5** Items to be kept at workshop

| |
|---|
| *Fast-moving items (to be ready at all times)* |
| oil filters |
| fuel filters |
| spark plugs |
| points |
| engine mountings |
| bulbs for headlamps, sidelamps, stop-lights, indicators |
| terminal and cabin mountings |
| fan belts |
| fuses |
| V-belts |
| spring bushes |
| wiper blades |
| ball-joints |
| gasket for rocker-arm cover |
| radiator hoses lower/upper |
| air cleaner element |
| speedometer cable |

*cont.*

shock absorbers
tyres and tubes

*Slow-moving items (to have few on stock)*

coils
injection nozzles
brake linings/pads
main leaf
U-bolts
centre bolts
water-pump repair kit
brushes for starter
brushes for alternator

*Consumption items (to have sufficient in stock)*

oils
grease
brake fluid
assorted nuts and bolts and washers
patches + solution gum
cleaning material
welding rods – gas and electric
gas
soldering fluid/powder
insulation tape
electrical connectors
electrical wire

A few items are to be kept on site so that the site mechanic can carry out the necessary daily maintenance and minor repair work.

**Table 4.6** Items to be kept at the field camp

oils and grease
brake fluid
assorted nuts and bolts and washers
patches + solution gum
cleaning material
insulation tape
electrical connectors
electrical wire

## PLANNING AND REPORTING SYSTEMS

Spare parts, tools and fuel are very expensive, so you must have total control of their consumption and use. Even if you have full confidence in your staff there is still a need to have a tight control system on the use of your equipment, as equipment and plant are very sensitive items. For example, a badly-tuned injection pump of a tractor may cause a high fuel consumption. If you do not have a daily fuel consumption record it might take a while before you detect this high consumption. This could cost you as much per day as employing 5 or 10 additional labourers.

Your planning and reporting system does not have to be sophisticated, but it should give you some information on the use of the equipment and the mechanical work carried out including the use of spare parts and consumables. For each piece of equipment or vehicle you need to know:

- operating costs over a certain time
- availability and utilization
- fuel consumption
- spare-part consumption
- its present condition.

The easiest system for a contractor is to have a separate log-book for each piece of equipment and vehicle. This can be a simple notebook which you can prepare yourself. The following details should be recorded on a daily basis:

- date
- working site or journey details
- odometer or hour-meter at the beginning and at the end of each day, plus distance or time driven
- fuel and oil issued
- signature of driver
  (see example in Table 4.7)

**Table 4.7** Example of a log book (operations)

| Date | Details of Journey | Km/Hr before | Km/Hr after | Total Km/Hr this Journey | Fuel Issued | Signature of Driver |
|---|---|---|---|---|---|---|
| | | | | | | |
| | | | | | | |
| | | | | | | |

On a separate page of the log-book, service and repair details could be recorded. You could also specify here when the routine services are planned, so that the driver/operator, site supervisor and the mechanic can always check when the next service is due.

**Table 4.8** Example of a log-book (maintenance)

| Date | Details of Repair/ Service | Spare Parts used | Consumable Items used | Signature of Mechanic | Next Service Due |
|---|---|---|---|---|---|
| | | | | | |
| | | | | | |

The daily movements of the equipment or vehicle should always be signed by the driver, while the repairs and services should be specified and signed by the person responsible for the repair (driver, mechanic, etc.)

It is important that you analyse the entries in the log books at least once a month. Take your time to do this; you might lose a lot of money if you don't!

Be sure to calculate the average consumption of fuel for each piece of equipment and record it. Analyse the data carefully and compare it with the data provided by the supplier. If the daily fuel consumption constantly remains too high there could be a problem with the engine and you should ask a mechanic to check it. But if the average consumption changes considerably from day to day, say 15 litres/100km on one day and 24 litres/100km on the next day, there could be something wrong with the speedometer, fuel could be siphoned from the vehicle, or the records might not be properly kept.

## SERVICE SCHEDULES

To keep equipment in good condition it is important to have it inspected and serviced regularly. You therefore have to plan the services accordingly and make sure the necessary arrangements are made in time. It is recommended to prepare a yearly service schedule for all vehicles and equipment.

Maintenance service schedules are based on kilometres driven or engine-hours operated. In addition, daily and weekly checks have to be carried out. However, it may be better to plan the service schedule on a regular time schedule, e.g. every two months, regardless of whether each piece of equipment has reached the targeted hours or kilometres (of course, you must not let the equipment pass the kilometre- or hour-based service limit, as set by the manufacturer; then you must undertake your regular checks more often). This system allows for a better planning of operations and gives your machinery a regular service in good time, regardless of whether the milometer or hour-meter stops working. Another advantage is that your staff will get used to regular services so that it becomes a routine. Ask your equipment and vehicle dealer for advice on the detailed requirements for services.

When planning the service schedule you have to consider:

- co-ordination with field operations (plan your activities in a way that the equipment/vehicles are available for the regular service without bringing your sites to a standstill)

- availability of the necessary spares and fuel, oil and lubricants (make sure all required resources are bought in advance and are ready when the service is due).

The service schedule should be prepared in such a way that the field operations are not disrupted. For example, all gravelling equipment from a particular site should not be serviced on the same day, but one by one so that the operations can go on without interruption.

# Hand Tools for Routine Maintenance and Regravelling

In labour-based road construction and maintenance, hand tools operated by labour are used to produce the same results as heavy bulldozers and graders do in equipment-based works. It is therefore obvious that it is important to select and maintain your tools properly.

This section describes the most common tools and gives information on how to use and maintain them. All your supervisors must be aware of this, in order to be able to instruct the labourers correctly.

Even if you are not in a position to procure on the local market hand tools that exactly follow the ideal standards described below, it is important to know the selection criteria. The choice of correct and good quality hand tools allows higher productivity while cheap hand tools will, although saving some money initially, reduce your productivity in the long run which could lead to lower potential profit on your contract.

It is also worth considering that the overall cost of hand tools is small compared to the labour cost. The cost of a shovel often equals something like five days' wages of a labourer. If you assume that a good quality tool lasts for 6 months heavy duty work, the cost represents only 3–4 percent of the labour cost. However, the condition of your hand tools strongly influences the productivity your labourers can achieve. If the labourers need to use tools that are not suitable, or are worn down, they cannot produce as much as they could with tools in good condition (in many cases you have only half the productivity if the tools are in bad condition). Remembering that your profit depends on your productivity, it is easy to see the strong economic argument for ensuring that your labourers have access to appropriate quality tools.

Spending some extra money when buying high quality tools usually pays off, because:

- low quality tools need to be replaced earlier than high quality tools
- working with worn or partly broken tools means low labour-productivity.

*Example:*
*You have a choice between two shovels, A. a robust model designed according to the specifications in this chapter at a price of 100 NU for 10 shovels, and B. with a handle that is a bit too short and a neck that seems weaker. In addition, the steel used for the blade of shovel B seems to be of lower quality than shovel A's. However, the dealer is offering you a very good price for shovel B, 75 NU for 10 shovels.*

*As shovel B is of lower quality (neck is weaker and lower steel quality), it will need to be replaced earlier than A; we can assume every 4 months instead of every 6 months. What is the cost per year? You need to buy 2 sets of shovel A per year or 3 sets of shovel B.*

*2 × 100 NU = 200 NU*
*3 × 75 NU = 225 NU*

*Shovel B is actually more expensive per year (by 25 NU)*

*Already this comparison shows that A is more advantageous. However, there is an additional major drawback, resulting in significantly less money for you. Productivity is lower with tools that are of lower quality or not properly designed.*

*The handle of shovel B is a bit too short, and the steel quality of the blade is lower resulting in the edge of the blade quickly becoming blunt. This results in reduced productivity. A very conservative assumption would be a 5% reduction because of the short handle and another 5% reduction because your labourers have to work with blunt blades for longer periods (even if you sharpen them regularly).*

*Assume labourers are paid 2 NU/day and that they work approximately 240 days per year.*

*10 workers × 240 days × 2 NU/day = 4800 NU*
*10% (5+5) lower productivity means 10% less produced by these 10 men during a year. To compensate for the work lost, you need to hire more people to do the same job.*

*10% of 4800 NU = 480 NU*

*So in addition to the 25 NU shown before we also lose 480 NU of potential profit on our contract(s).*

The most common tools used on a labour-based road work site are:

pickaxe and mattock
shovel and spade
hoe and forked hoe
rake and spreader
sledge-hammer
crowbar
earth rammer
bow saw
bush knife
grass cutter
wheelbarrow

## PICKAXES AND MATTOCKS

*Specifications*

Pickaxes and mattocks are tools for excavation. Many different designs are produced for agricultural work. The most common pickaxe for road construction is shown in Figure 4.5 and a mattock is shown in Figure 4.6.

These tools always have an oval eye so that the handle cannot turn in the eye. Both the pickaxe and the mattock are rather heavy; the pickaxe usually weighs between 2.7 and 3.6kg and the mattock between 1.8 and 2.7kg. As they are double-edge striking tools it is necessary to have a straight handle with an

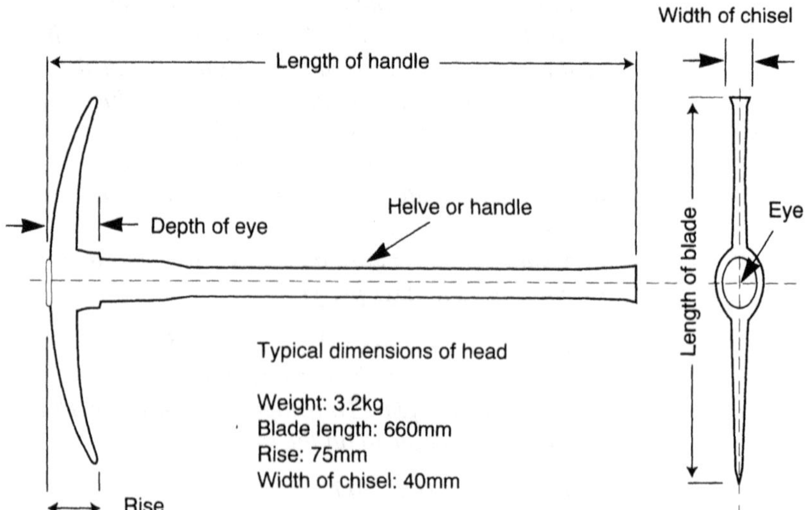


Figure 4.5 *Pickaxe*

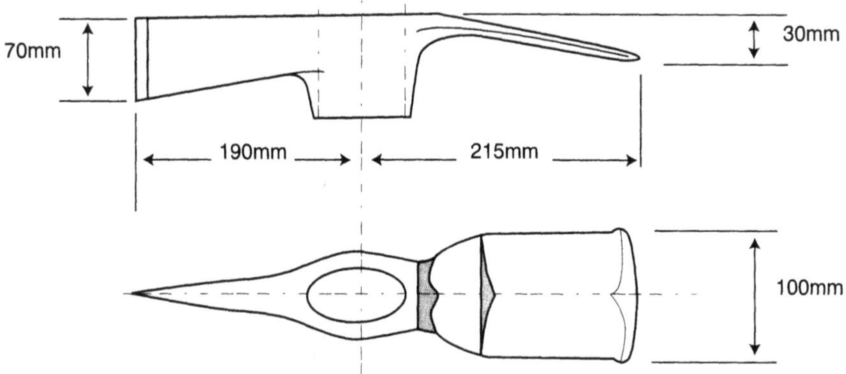


Figure 4.6 *Mattock*

elliptical rather than circular cross-section. The handle should also be provided with a raised safety grip which prevents the handle slipping out of the worker's hands. They are usually made of hickory or the nearest equivalent quality hardwood (see figure 4.7).

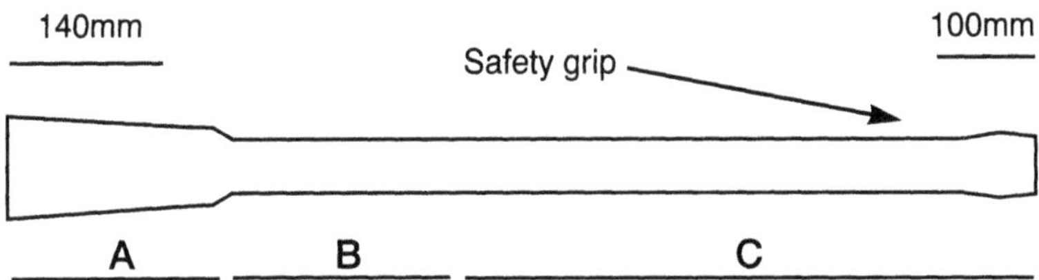


Figure 4.7 *Pickaxe and mattock handle*

*Maintenance and Repair*

Both these tools need sharpening from time to time. The mattock, being less solid, can be sharpened on a grinding stone, while the pickaxe may need to be re-forged from time to time. A local blacksmith may be hired for this work. Handles need to be replaced when broken. It is therefore advisable to have spare handles in the site store and to train the store issuer to replace handles correctly.

*Use*

The pickaxe is used to loosen stony material, mostly in quarries. It is also used for digging ditches in hard soil. The mattock is used to loosen firm soil which cannot be dug by a hoe, and also to cut roots. It is used in sloping the side drains because it has a wider cutting edge than the pickaxe.

## SHOVELS AND SPADES

*Specifications*

Shovels are used for scooping up material and throwing it on to a trailer, truck, wheelbarrow, or directly to where the material is needed. The shovel has a rounded or pointed blade while a spade has a square shaped blade which is stronger than that of a shovel. Spades can also be used to loosen soil.

The handle for both tools should be long enough to allow the worker to throw the soil with little effort. For workers with an average height a length of 65 to 70cm is recommended.

Shovels and spades should not have sharp reverts or joints which damage the hands of the user. When buying shovels or spades ensure that the blade/handle joint is smooth.

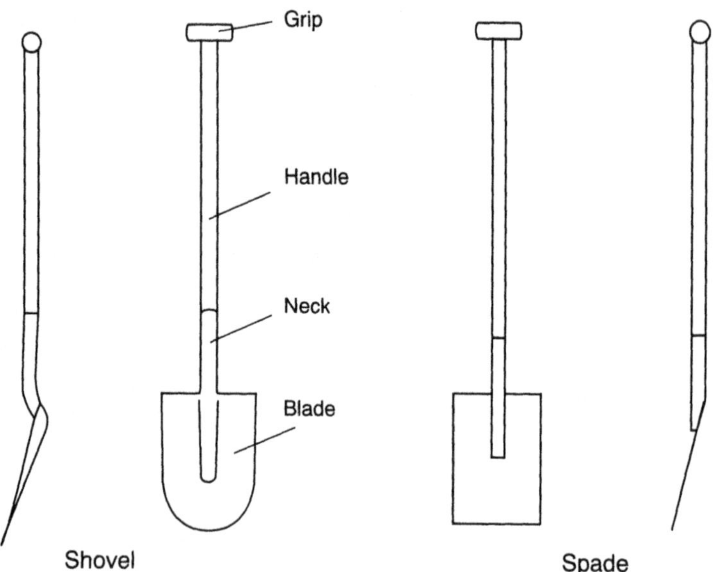


Figure 4.8 *Shovel and spade*

*Maintenance and Repair*

The blade of a good shovel will not bend or crack but it will, of course, wear with time. The edge of the blade will become blunt, making it difficult to push into the soil. The worn out blade can sometimes be cut and sharpened so that the shovel can be used again.

*Using Shovels and Spades*

When the soil is loose the shovel can be used directly to scoop it up and throw it elsewhere. With harder soil the spade is more useful, because it can be pushed into the ground without

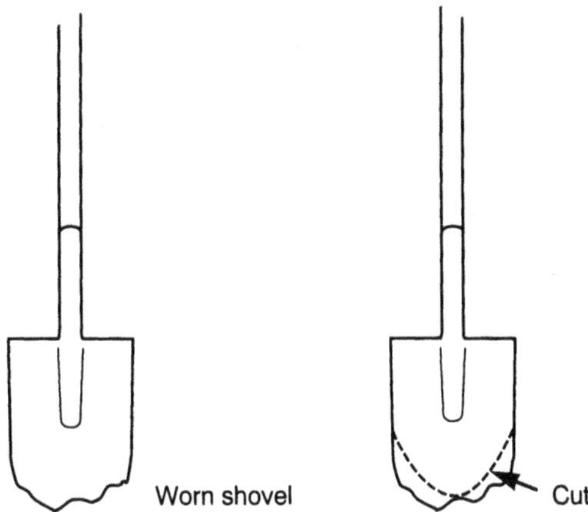


Figure 4.9 *Repair of shovel blade*

bending the blade. To help push the blade into the ground, the worker can put his foot on the top of the blade and press it down. To be able to do this the worker should wear shoes with hard soles.

## HOES AND FORKED HOES

*Specifications*

*Hoes*

The hoe is, in addition to being very useful in agriculture, also an important tool in labour-based road works. In road work it is used for tasks such as loosening soft soil, shaping of slopes and grubbing. It consists of a blade and a handle. The eye can be

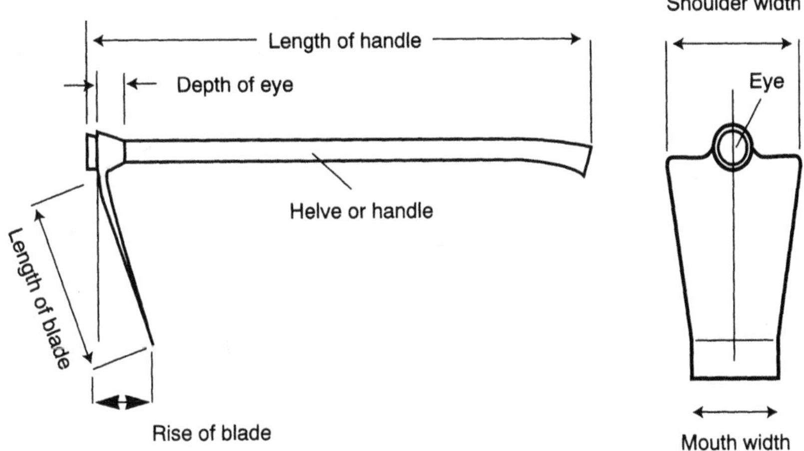


Figure 4.10 *Hoe*

round or oval, although for road works the oval eye is recommended. The round eye makes it easier to replace the handle but you cannot hold the handle as tightly as the handle tends to turn. The blade of the common hoe has a straight cutting edge.

*Forked Hoes*
The forked hoe is a special type of hoe. Instead of a blade it has a number of prongs which can penetrate a cohesive or hard/stony soil easier than a blade.

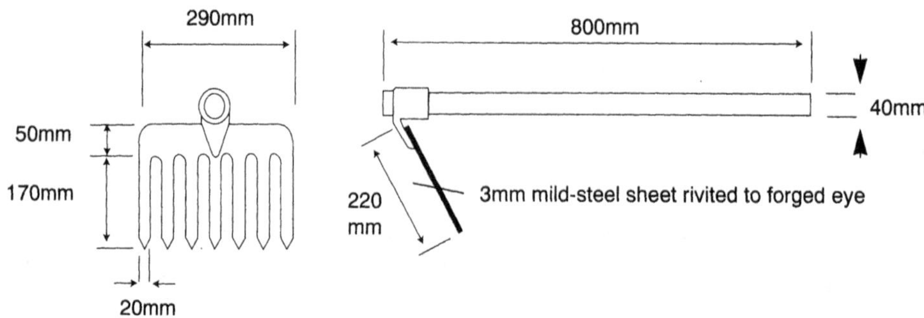

Figure 4.11 *Forked hoe*

*Maintenance and Repair*
Hoes must be kept sharp to be effective. The cutting edge should preferably be sharpened on a grinding wheel, but a flat metal file can also be used.

The prongs of the forked hoe become blunt after a while and they need to be sharpened on the inner face. Again, a grinding wheel or flat iron file can be used for sharpening.

The site store issuer can be asked to carry out such work when not busy with issuing or replacing tools.

Figure 4.12 shows how the edges of hoe blades and prongs should be sharpened.

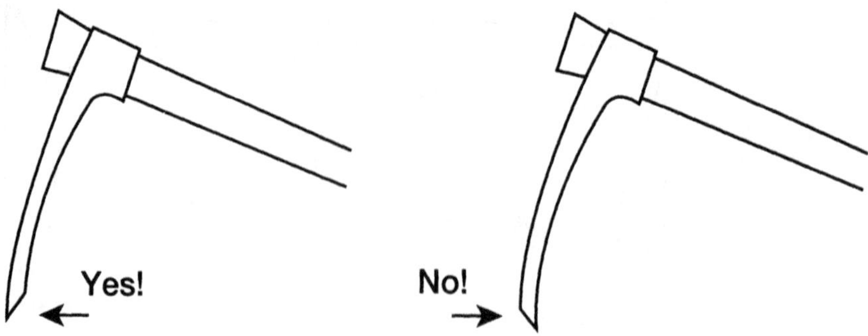

Figure 4.12 *Sharpening of hoes*

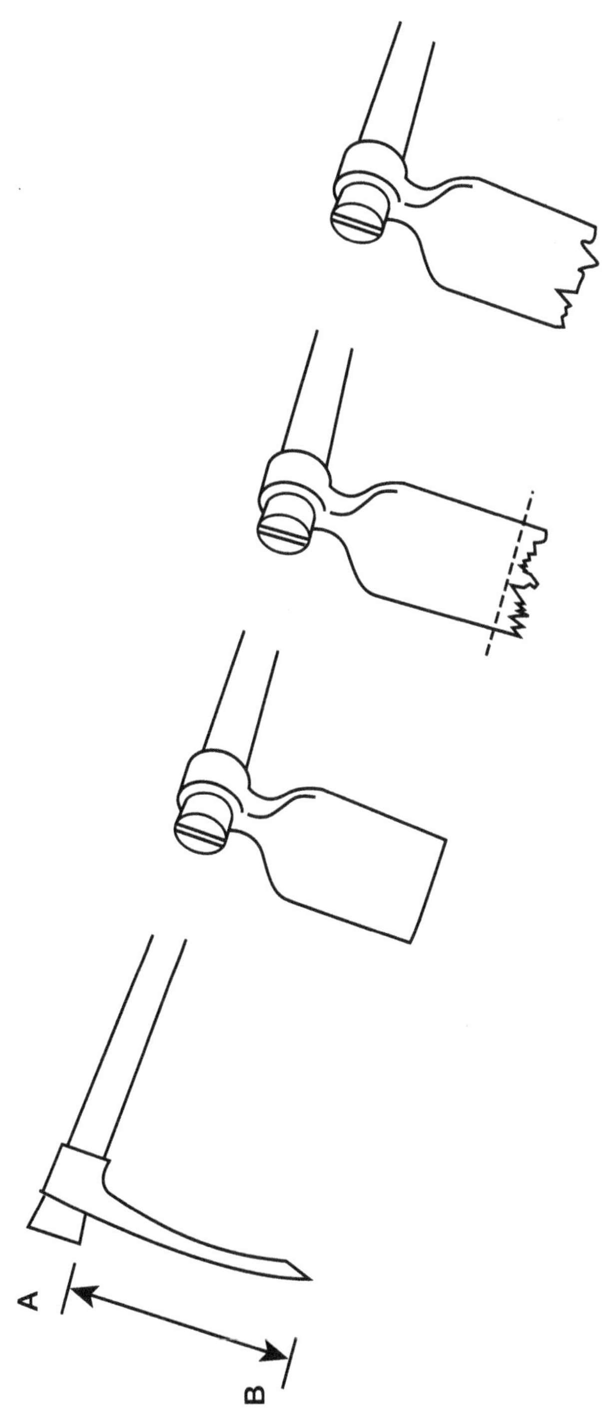


Figure 4.13 *Repair of hoe blades*

If the blade has been chipped or pieces broken off, the hoe should not be used until it has been repaired. The repair work can be done by cutting or filing the edge straight and then sharpening it (see Figure 4.13). When the length of the blade (A–B) is less than 15cm, the hoe is no longer efficient for digging but may still be used for agricultural activities like weeding.

Handles must be well dried before they are shaped and fitted. One end of the handle should be bigger than the other so that it can be firmly fixed into the eye. The length of the handle should be such that it is comfortable for most workers to use.

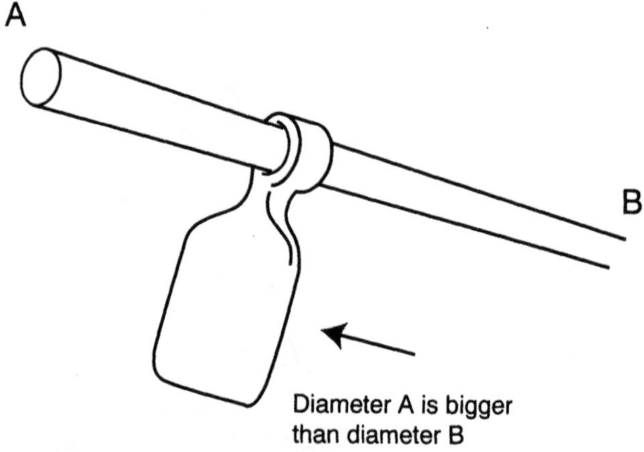


Figure 4.14 *Handle for hoe*

It is recommended to use wedges to fix the handle effectively to the blade, as shown in Figure 4.15.

*Using Hoes and Forked Hoes*

Hoes can be effective when:

- grubbing and removing topsoil
- loading baskets or trays
- loosening soft soils
- shaping slopes
- spreading

Forked hoes can be effective when:

- excavating hard and stony soils

In excavation the plain hoe is best for rather soft soils with few or no stones, while forked hoes are better on cohesive or hard and more stony soils.

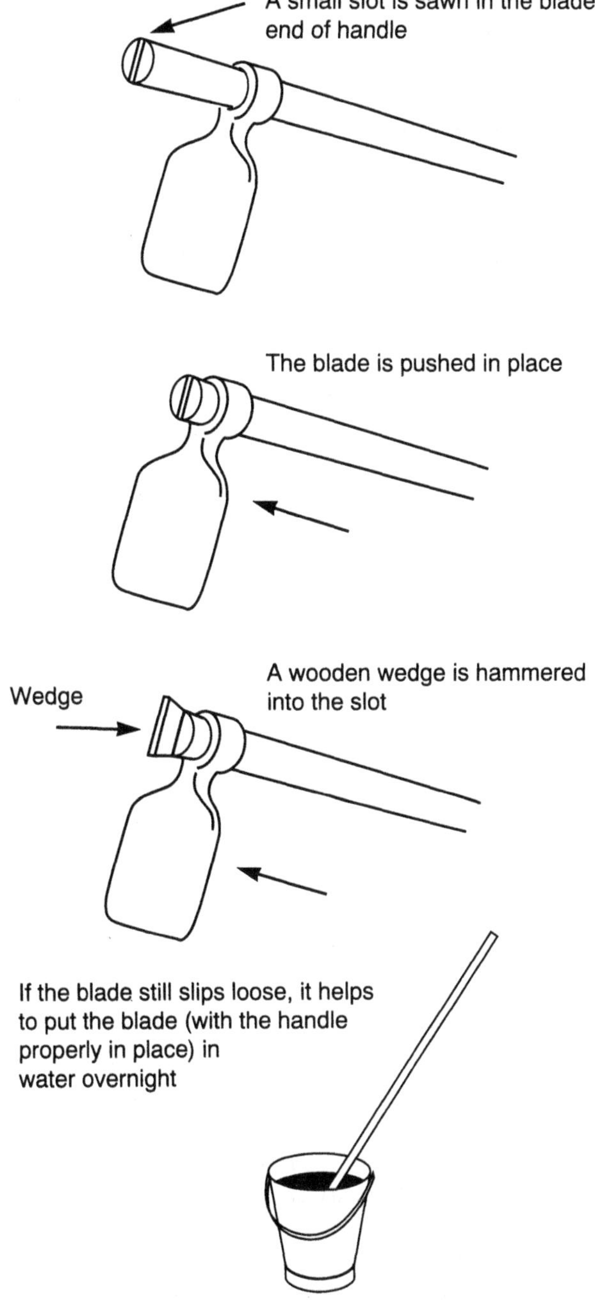


Figure 4.15 *Fixing handles*

## RAKES AND SPREADERS

*Specifications*

There are numerous variations of rakes and spreaders, all designed for specific purposes. Rakes are used in road works for raking out vegetation from loose soil. Commercially-produced rakes have 10 to 16 teeth, each about 75 – 100mm long, with an overall width of about 400 – 450mm. They require straight handles made of hard wood or metal tubes.

Spreaders are used for spreading out the soil on fills. A spreader can be a heavy-duty rake but the best spreaders are specially made for its purpose. They can be made of sheet metal (3–4mm thickness) and have a ridge for crushing lumps of soil.

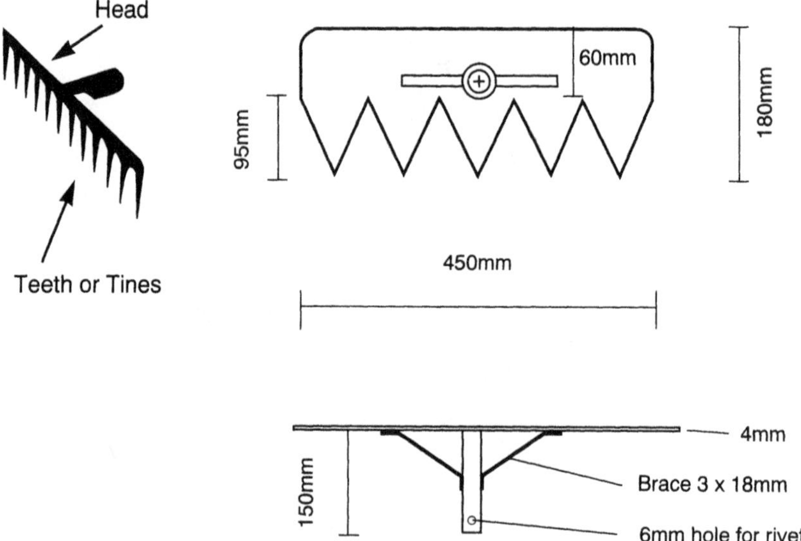


Figure 4.16 *Rake and heavy duty spreader*

*Maintenance and Repair*

Apart from replacing broken handles, spreaders require no routine maintenance. Cracks in the blade can be welded. Broken rakes can also be welded. If a rake tends to break often, it can be reinforced with a piece of round iron bar which is welded across the upper part of the teeth.

*Using Rakes and Spreaders*

The rake is used for collecting vegetation from loose soil when grubbing, but can also be used for spreading if the soil is not stony. The spreader is very useful when forming the camber and for spreading gravel. The soil or gravel should be raked from the centre line towards the shoulder.

## SLEDGE-HAMMERS

*Specifications*

The sledge-hammer most useful for road work is the double-faced head hammer. The eye is oval shaped as shown in Figure 4.17.

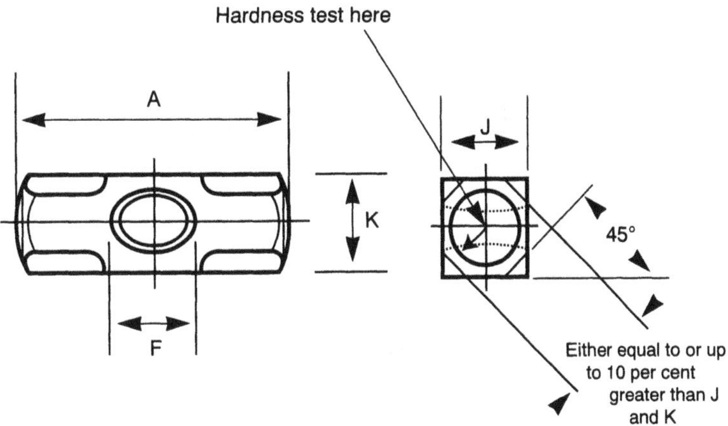


Figure 4.17 *Sledge-hammer*

For a double-faced striking tool it is necessary to have a straight handle with an elliptical, rather than circular, cross-section. The handle should also be provided with a raised safety grip at the end.

*Maintenance and Repair*

Apart from replacing broken handles, sledge-hammers do not require routine maintenance.

*Using the Sledge-hammer*

Sledge-hammers are used for breaking large boulders or rock. The worker should always wear goggles to protect the eyes.

## CROWBAR

*Specifications*

The crowbar is, like the pickaxe, used mostly in stony or very hard soils. The crowbar looks like a simple tool but it has to be of a very strong material that does not bend easily and must be well designed to function properly. They are usually manufactured either as round or octagonal section rods. For road construction the diameter should be over 30mm to give a good, firm grip. The length is required to be between 1.5 and 1.8m. Reinforcement bars cannot be used for crowbars as they are too soft.

The crowbar has a hammer end and is pointed or chiselled at the other. A pointed end is perhaps the most common. The chisel end is more useful for leverage.

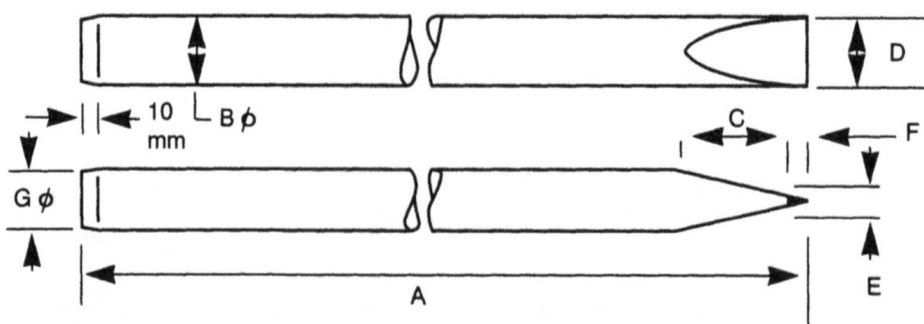


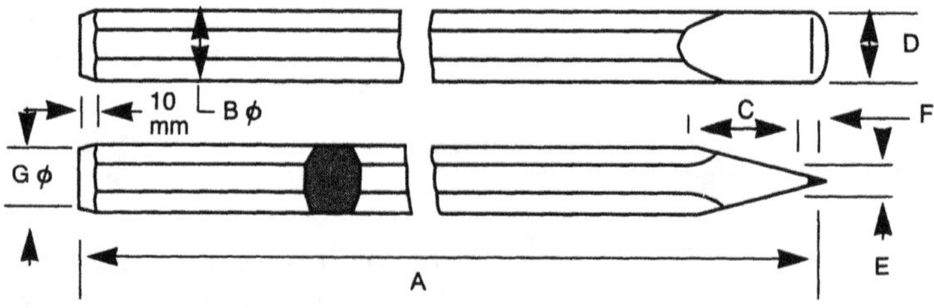


Figure 4.18 *Crowbars*

*Maintenance and Repair*

A crowbar requires little maintenance and the ends only need to be pointed or sharpened again.

*Using the Crowbar*

The crowbar is mostly used for breaking loose hard material, or moving boulders or heavy things when used in the right way as a lever.

## EARTH RAMMERS

*Specifications*

The earth rammer is used for compacting soil, and consists of a weight with a handle. Two aspects determine the effectiveness of an earth rammer: its weight and the area that hits the ground. Ideally the weight should be as large as possible and the area as small as possible. A rammer which can be handled by a worker should therefore have a weight of some 8 – 10kg and a diameter of the bottom end of 13 - 15cm. The handle must be long enough to allow the worker to lift the rammer without bending his back.

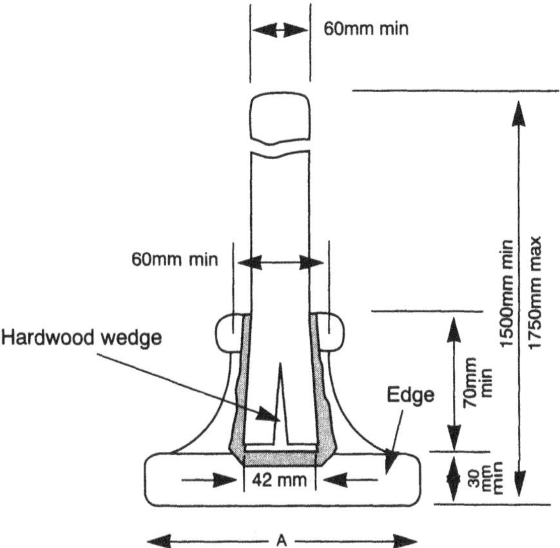


Figure 4.19 *Earth rammer*

*Maintenance and Repair*

If the handle is of wood it will have to be replaced regularly. When fitting the wooden handle, a wedge is placed in the centre of the handle and hammered into the rammer. This makes the handle edge flare up inside the metal rammer and prevents the handle from coming out.

*Using Earth Rammers*

Rammers should be used for small compaction work where mechanical compaction is not available or not possible, e.g. backfill of culverts or structures, trenches, pot-hole filling, etc. The soil should be compacted in thin layers of not more than 10cm.

## BOW SAWS

*Specifications*

Steel framed bow saws are commonly used for cutting small trees and branches. A relatively narrow blade is held in tension by the frame. Tension is applied by a quick release lever. The lever combined with an oval sectioned frame, provides a comfortable hand grip. Blades are 20 – 25mm wide and are produced in a standard length. They can be supplied with various shaped teeth to suit the different types of wood. The frame is made of mild steel and the blade is made of high-carbon alloy steels.

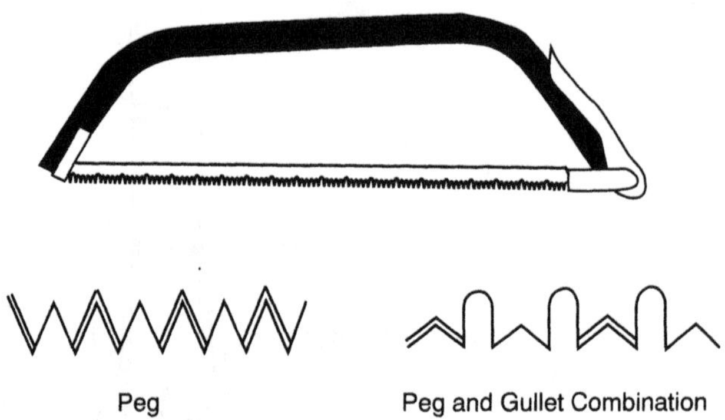


Figure 4.20 *Bow saw*

*Maintenance and Repair*
No maintenance is required except to replace the blades. It is advisable to open the tension lever if the saw is not used for long periods. This helps to retain the original tension of the blade.

*Using Bow Saws*
The bow saw is used for cutting down small trees and branches. It is also used for any woodwork that may be needed at site, e.g. cutting planks, pegs, etc.

## GRASS CUTTER

*Specifications*
The grass cutter consists of a metal strip (mild steel) 50mm wide, 3mm thick and about 100cm long. The bottom 20cm is cranked and sharpened on both edges, and the top end is fitted with a wooden shaped handle riveted on to the metal strip.

*Maintenance and Repair*
When the blade gets blunt it has to be sharpened using whetstones, files or grinding wheels.

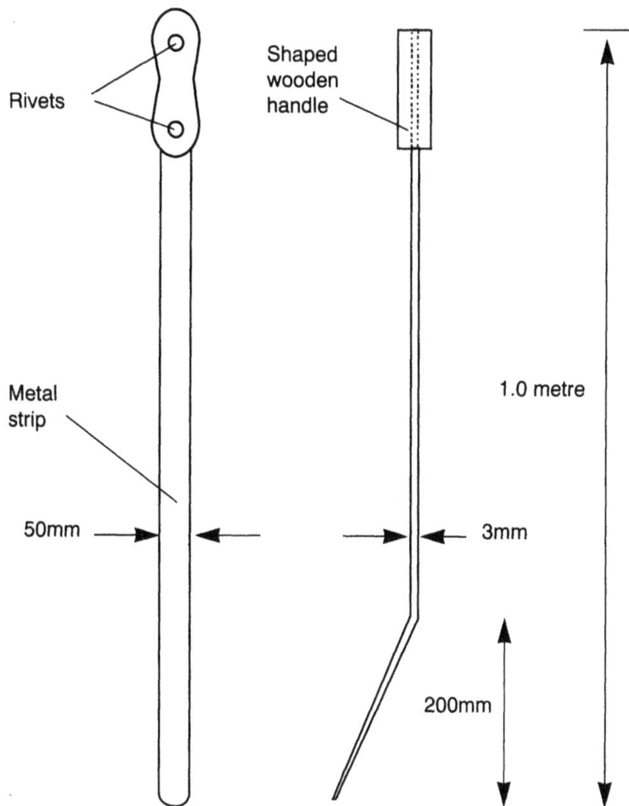


Figure 4.21 *Grass cutter*

*Using Grass Cutters*
It is used for grass cutting, especially for routine maintenance of the road reserve.

## WHEELBARROW

*Specifications*
The wheelbarrow is one of the most useful and economic forms of transportation equipment over short distances (up to 200m). Wheelbarrows can be of many different types and qualities. A good wheelbarrow should take a big load (struck capacity approximately 60 to 70 litres) and be easy to balance and tip.

A wheelbarrow consists of a body or tray which rests on a chassis with attached handles and a wheel. It also has a stand. The strongest and most comfortable wheelbarrows have rubber wheels and a tray made of 1.6 mm to 2.0 mm steel sheets. The tray should be reinforced around the rim and attached to the chassis with bolts, nuts and washers.

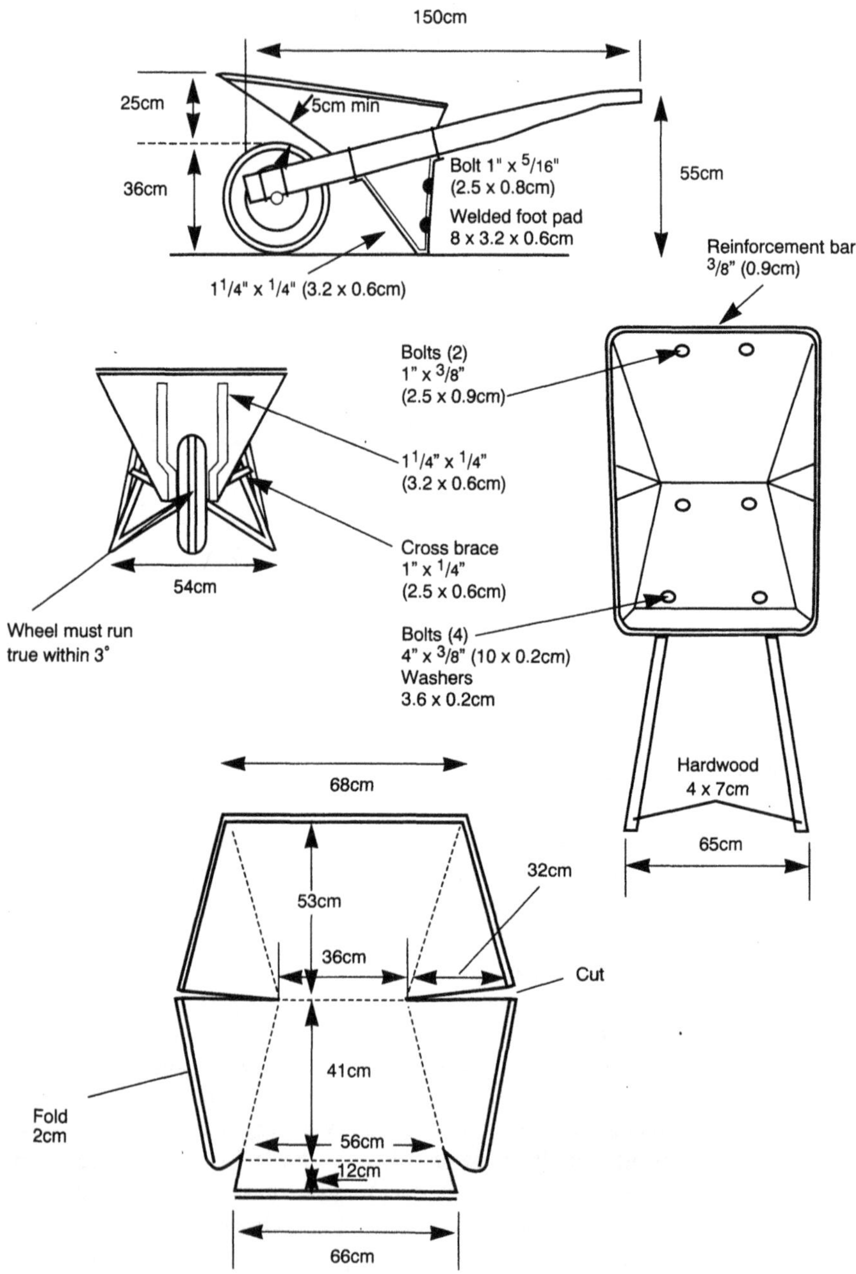


Figure 4.22 *Wheelbarrow*

*Maintenance and Repair*

Wheelbarrows require frequent maintenance to remain serviceable. Each day all bolts and nuts should be checked and tightened or replaced if necessary. The bearing of the wheel should be greased from time to time.

*Using Wheelbarrows*

Wheelbarrows are used to transport material. Where several wheelbarrows are used to transport excavated soil, it is necessary to organize the work so that they do not get in each other's way. Special attention has to be given to the dumping of material, to avoid double handling.

# CHAPTER 5: INTRODUCTION TO LABOUR-BASED ROAD CONSTRUCTION

## Learning Objectives

After you have read this chapter, you should be familiar with the following issues related to major road construction operations:

- site organization and support work
- construction activities
- setting out, horizontal and vertical alignment
- drainage and erosion control
- work productivity
- low-cost structures and culverts
- gravelling
- alternative pavements
- planning and reporting systems.

## Introduction

This chapter will provide you with an introduction to labour-based road work methods. Although you, as a regravelling or maintenance contractor, will probably not be involved in road building, it is necessary for you to understand the principles of these methods. From the expression 'labour-based', you know you will be working with labour rather than equipment. Handling large groups of labourers is usually more of a management challenge than to co-ordinate equipment-based work. An efficient work organization with good planning, monitoring and supervisory systems are the basic requirements for running a site successfully.

When undertaking maintenance contracts, you may be asked to undertake repair or reconstruction. As always, the construction details relevant to your contract will be specified in the contract specifications. This chapter helps you to understand these specifications and carry out your work correctly.

# Site Organization and Support Work

## ORGANIZATION AND STAFFING

A labour-based road construction project, with as many as 150–200 labourers, and a work-site that is several kilometres long, needs to be planned and organized carefully. Otherwise it is impossible to ensure high productivity from each labour gang and an efficient use of each piece of equipment. To be able to control the work, it must be split into simple operations. Each operation is then usually assigned to a separate labour-gang with its own gangleader in charge. To check that the activities of each gang meet the targets of the overall construction schedule, there is usually a site supervisor co-ordinating the gangs. Figure 5.1 shows an example of how a labour-based road construction site can be organized.

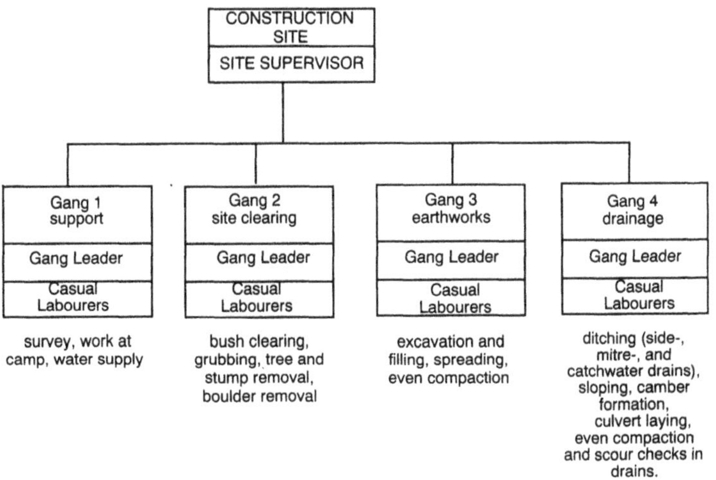


Figure 5.1 *Typical labour-based construction organization*

A construction site supervisor is usually responsible for a total labour force of some 50 to 100 workers. Training should be provided for new supervisors, as good management skills are needed to control such labour forces.

For example, experienced supervisors advise that it is best to allocate work according to type of task. That allows gangs to become specialized, which normally results in higher productivity.

## SITE CAMP ESTABLISHMENT

Before any site work can start, a number of preparatory activities have to be carried out. Technical and operational planning work, procurement of the required material and, on remote sites, the establishment of a site camp.

When establishing a site camp for a labour-based roads project you should consider these major points:

*Location:*
When you undertake a site inspection before preparing the bid, think carefully about the location of the site camp. Here is a check-list to go through before deciding on the location:

- Camps should be within walking distance of all works. The distance from the camp to the furthest working place should not be more than 4km. For roads longer than 8km, you will need to shift the camp.
- Camps should be located on high, well-drained ground.
- Camps should be close to a water source.
- Camps for major structural work should be next to the construction site to avoid hauling construction materials over a long distance.
- Camps should preferably be located away from villages and market centres but not so far as to cause unnecessary problems with the purchase of food, etc.
- Camps should be accessible to project vehicles.

*Infrastructure*
The size of the camp depends on what kind of work you are undertaking and how many people need to be housed in the field. Movable field huts, manufactured locally, provide the comfort and security that are needed if your workers are to achieve the levels of productivity which you expect (tents do not provide enough security).

Standard requirements for a construction site field camp are:

- one field hut for the site supervisor
- one field hut for the site office
- one site store
- toilet and bathroom hut
- field huts for housing of the plant operators during gravelling (if equipment is used)
- extra site store for fuel, oil and lubricants (for gravelling only).

*Timing*
The camp installation is normally the first activity on a road construction project. Labourers must be called to fence the compound, to excavate pit latrines and to assist in erecting the field huts. It is usually better to have the camp fully established before actual construction work starts.

*Staff*
Apart from the site supervisor, it is often useful to have a store-keeper to keep track of materials and tools. Besides reducing waste and ensuring that material is available when needed, a store-keeper should also be able to carry out minor repairs to tools.

Depending on the local security situation, it might be necessary to hire watchmen to guard the compound.

If the camp is not very close to a water source, water carriers must be employed. It is important that the labourers on site are provided with clean drinking water.

For gravelling sites, where the hauling of material is done with tractor/trailer combinations or lorries, the operators and drivers are also based at the site camp and require accommodation.

When planning your site camp it is useful to try out alternative solutions on sketches you prepare yourself. Your sketches should show all the different installations and the transport facilities to make sure that movement of goods and labour will run as smoothly as possible. When the final decision is made, this sketch should be given to the site supervisor. It is then much easier for him or her to prepare the camp.

## SITE STORES, EQUIPMENT AND TOOLS

To reduce theft and excessive wear and losses of tools and equipment, each site should have a field store either in the form of a hut or at least in lockable tool boxes. Whatever storage method is used, the equipment and tools should be arranged in an orderly manner and the store must be clean, dry and protected from weather and theft.

The number and types of tools and equipment required depends, of course, on the task and the number of workers employed. A proposed standard list of tools and equipment is presented in Table 5.1. The amount shown would be sufficient for 80 to 100 labourers on a construction site. This list can be used as a general check-list. You might need to change or supplement it for a specific project.

**Table 5.1** Standard list of site tools and equipment for construction

| *Description* | *Number required for construction work* |
|---|---|
| Hoes | 70–100 |
| Forked hoes | 30–80 |
| Shovels | 50–80 |
| Mattocks | 20–40 |
| Pickaxes | 30–40 |
| Bush knives | 10–20 |
| Slashers | 20–30 |
| Spreaders | 10–20 |
| Garden rakes | 10–20 |
| Axes | 5–10 |
| Sledge-hammers | 5–10 |
| Mason hammers | 5–10 |
| Earth rammers | 5–10 |
| Flat files | 10–20 |
| Wheelbarrows | 5–30 |
| Claw hammers | 2–5 |
| Crow bars | 5–10 |
| Spirit-levels | 5 |
| Tape measures | 4 |
| Ranging rods | 10–20 |
| Boning rods | 2 sets |
| Ditch templates | 3–5 |
| Ditch-slope templates | 3–5 |
| Camber boards | 1–2 |
| Straight edges | 2–5 |
| Strings | – |
| Ropes | – |
| Water buckets | 8–20 |
| Bushman sows | 2–5 |
| Mason tool-kits | 1–2 |
| Grinding wheel | 1 |
| Anvil and pair of bellows | 1 |
| Nails | – |
| Water containers, 50 gallons | 2–5 |
| First aid kit | 1 |
| Levelling instrument (Abney level, line & level, etc.) | 1 |
| Compaction equipment | 1 |

Note:

- Light compaction equipment can be tractor/manual or animal drawn. Also pedestrian vibrating equipment can be used.

- In rocky terrain tools like forked hoes, pickaxes, sledge hammers, chisels and crowbars must be available in sufficient numbers.
- Earth rammers are mainly used to compact the back fill when installing culverts.
- The required standards for handtools are described in Chapter 4.

## Employment of Casual Labour

On a building site you usually employ your own relatively small staff of artisans and labourers that you have trained on the job. They form a regular team who shift from one site to another. If you require additional labourers you probably employ a small number of local people to work with your regular team.

In labour-based road construction, large labour forces have to be employed. As it would be far too expensive to have a permanent team, labourers have to be employed on a temporary basis from within the vicinity of the road. To avoid the problems and costs of setting up a site camp, they should be able to walk from their homes to the site. On longer roads you may even have to employ more than one labour force, as the walking distance would become too long. In addition, importing labour from other areas is often not acceptable to the local community.

The employment process is a major management task that must not be neglected by a labour-based road contractor. The contractor needs to ensure that his employment conditions follow the rules set in the labour legislation of the country and, as the number of labourers to be hired is high, following a set of simple and uniform procedures enables you to manage the employment process in a fair and efficient way. This will be looked at in detail in Chapter 8.

## Construction Activities

After the planning and preparation works are completed, and the site camp is established, the actual construction work can start.

The separate operations on a construction site have to follow each other in a logical sequence. The construction process should be broken down into small and simple activities. Distribution of labourers working on each operation has to be carefully assessed, and sufficient time has to be allocated for the transfer of labour from one activity to another.

To make planning and controlling the site works easier to manage, each operation is broken down in a number of practical and logical activities. These activities must follow each other in defined sequences.

Table 5.2 gives a general view of the normal construction sequences, although they may differ from project to project:

**Table 5.2** Construction operations and activities

| *Operation* | *Activity* |
|---|---|
| Support | setting out alignment<br>work at camp<br>water supply<br>erosion protection |
| Site clearing | setting out of detailed activities<br>bush clearing<br>grubbing<br>tree and stump removal<br>boulder removal |
| Earth work | excavation and filling<br>spreading<br>(compaction) |
| Drainage | ditching (side drains, mitre drains, catch water drains)<br>sloping and back sloping<br>camber formation<br>(final compaction)<br>culvert installation<br>ditch erosion controls |
| Structures | multiple line culverts, drifts, bridges (large structure work to be carried out at a separate site with specialized personnel) |
| Improvement of quarry access road (if necessary) | improvement work to allow equipment to pass. (continuous process during gravelling work) |
| Quarry preparation | opening of quarry (removing of trees, crops, etc. removing overburden)<br>excavation and stockpiling of gravel |
| Road preparation | re-shaping of road according to the desired standards<br>carrying out small repair work, e.g. washouts, culvert replacement |
| Quarry work | excavation and stockpiling of gravel |
| Hauling | hauling of gravel from quarry to road |
| Road work | re-shaping of road (if not already done)<br>off-loading of gravel<br>spreading and compaction<br>(provision of gravel stocks for maintenance) |

Trained supervisors, responsible for a site, are usually capable of controlling effectively a labour force of from 50 to 80 workers. Gangs, formed for the different operations, range from 10 to 25 labourers, according to the nature and amount of work to be carried out. Balancing of the gang sizes, i.e. making sure that the labour is used in the most efficient way, is a daily task of the site supervisor. It is determined by:

- the amount of work to be done for each activity
- how much can the labourers produce per day on each task (task-rates)
- how many labourers are available on site
- how the activities follow one another.

A supervisor must always plan ahead by at least one day. After the labourers have completed their daily work the supervisor must record the output achieved on all activities. Based on the production achieved and the overall plan for the project, a plan for the following day has to be prepared. In this plan the daily production targets for the activities are set.

To prepare these work plans properly, the supervisor needs to know what has happened on the site before. Without information such as what resources were needed to produce a given output, why certain targets were not met, and so on, proper planning is impossible. To get the right information on time, a well-functioning reporting system is needed.

There is, however, a problem concerning all site reporting systems. On the one hand you want them to be sufficiently detailed to give all the information needed. One the other hand it is not realistic to expect site staff to fill in reports for several hours every day after the labourers have left. You need to settle for a compromise where you collect only the information that is absolutely necessary and that your forms are as simple as possible.

Remember, if you have a system where there is a reward for keeping proper records, the result is much more likely to be provided on time and in a correct way. (See also the section on Planning and Reporting Systems in this chapter.)

At the end of this chapter a typical example of how construction work is divided into simple activities is shown.

# Setting Out: Horizontal and Vertical Alignment

## GENERAL

The surveying requirements for labour-based road construction vary with the type of work to be executed. The construction of new roads requires a complete survey to establish the best possible alignment. The improvement of existing roads generally requires only some minor corrections.

The establishment of the alignment is a task for an engineer of the client and must already have been done during the technical assessment of a road. The choice of alignment has great implications for costs and construction methods.

The detailed surveying (e.g. establishment of curves, cut and fill, drainage, etc.) is done shortly before construction starts. This is undertaken by the site supervisor under the guidance of the client's engineer, using basic data obtained during the assessment stage.

It is extremely important to set out the centre line of a new road well in advance of the start of the construction works. This will allow the sorting out of right-of-way problems with the local people, and will also ensure that no new crops are planted in the road area.

The surveying should be done using the simplest instruments and methods possible. The following list shows the elementary aids and instruments used for surveying and setting out:

Pegs (reference, survey and multi-purpose)
Tape measure/measuring wire
Ranging rods
Boning rods
Profile boards
Slope templates
Ditch templates
Gradient templates
Camberboards
Straight-edge in combination with spirit level
Line and level in combination with ranging rods
Abney level
Dumpy level

It is not possible to describe all of these measuring and control aids in this chapter. However, the most useful instruments and their practical use for the type of work you as a regravelling and maintenance contractor undertake, are explained.

**TAPE MEASURE**

A great variety of tape measures exist. The most common length of tape measure used for setting out is 30m. The tapes are made of steel or linen. Although the former is stronger, the numbers/marking on the tape become unreadable after a period of use.

Note: the zero point is not always located at the same place on different tape measures.

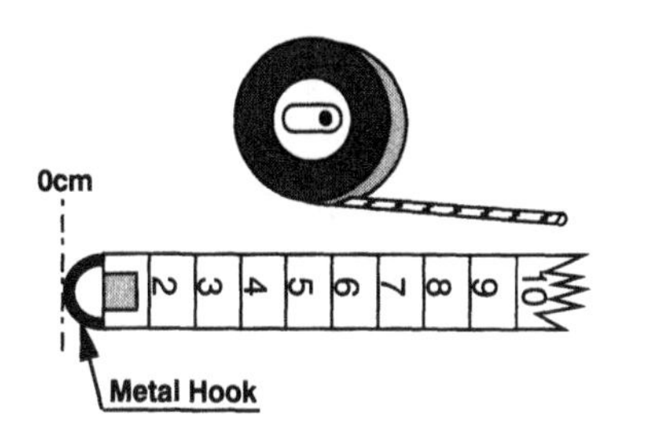


Figure 5.2

**RANGING ROD**

Ranging rods are sticks usually 2m long with a diameter of approximately 2.5cm. They are made of various materials (metal, hard plastic, wood) and are usually provided with a pointed metal end. They are painted alternately red and white with black marking at the 1 metre point. The length of the red/white sections is 50cm.

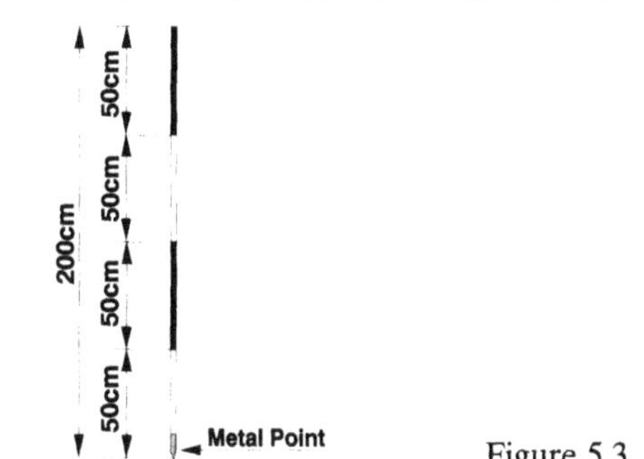


Figure 5.3

**PROFILE BOARD**

A profile board is designed in such a way that it can be attached to a ranging rod. It has a screw mechanism that enables the profile board to slide up and down on the ranging rod and be fixed at any desired point simply by tightening the screw. A long-lasting profile board is one made from thin steel plate (40cm x 10cm) welded to a short length of metal tubing that can slide up and down and can be clamped to the metal rod.

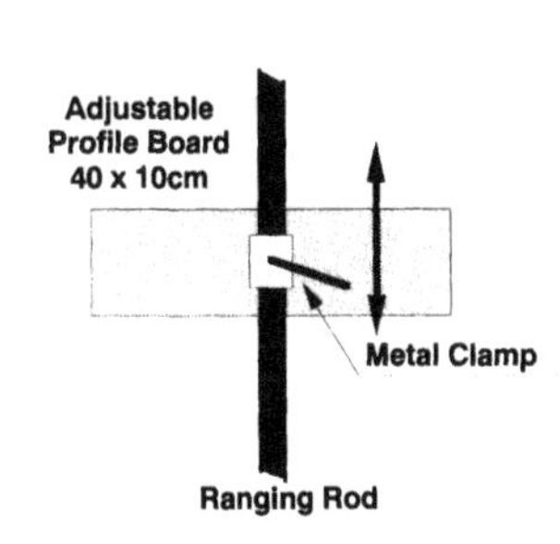


Figure 5.4

**BONING RODS**

Boning rods are T-shaped and of a uniform height. They can easily be manufactured by nailing a wooden lath of 80cm length and 10cm height on to another lath of 130cm length and 10cm width so that the end result looks like a 'T'. The horizontal lath should be painted in clearly visible colours. Boning rods have to be used in a set of three.

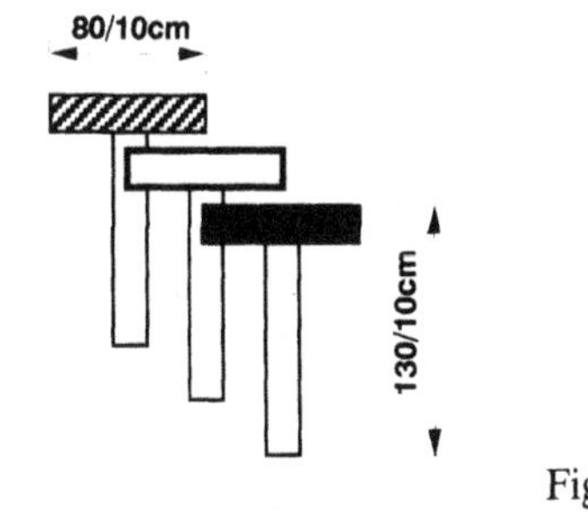


Figure 5.5

STRING AND LINE LEVEL

A line level is a small spirit-level about 80 – 120mm in length. It has a hook on each end which is used for hooking the level on to a smooth line. The level is used together with a string line, ranging rods (or profile boards) and a tape measure. The line level requires two people to operate.

Always check:

- that the line is smooth or of nylon
- that the line is tight
- the level is in the middle between the two ranging rods
- the accuracy of the level regularly.

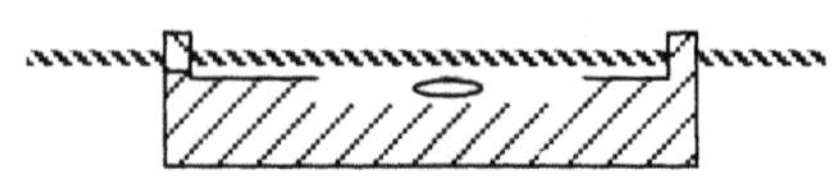

**to check the accuracy of the level:**

- place two ranging rods 20m apart
- fix a line on the 1m-mark on one rod and transfer the level to the other rod; mark this level
- keep line in place and turn the level round
- adjust line again and mark the new level and measure the difference between the two levels; if difference is less than 10cm the correct level is exactly in the middle of the two marks, if the difference is more than 10cm the level should be changed.

Figure 5.6

TEMPLATES

Templates are used to control certain shapes of the road. For example, to control the correct shape for the slope and ditch a template of the standard slope-ditch size can be used by the labourers to check continuously whether the correct shape is dug. Templates are usually made of wood and tailor-made for each particular project, in accordance with the standard measurements.

Templates are very useful control aids as any labourer can see the exact size and shape of the work she or he is required to carry out.

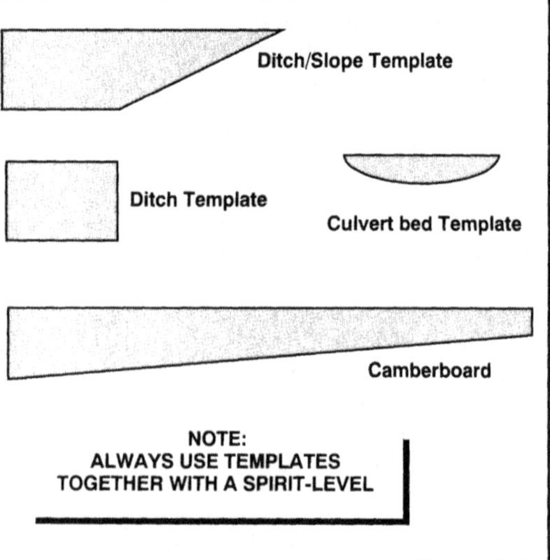


Figure 5.7

## CONTROL OF ROAD ALIGNMENT

In this section we shall discuss how the measuring aids described above are used for road alignment control. You may need these methods later when it actually comes to carrying out regravelling or maintenance work for different control functions, or to re-establish levels and directions. There are different methods used in the various projects and you therefore need to refer to the specific project methods when it comes to actual implementation. However, most setting-out techniques should be basic knowledge for a labour-based road contractor and his supervisory staff.

*Setting out a straight line*

For several activities straight lines must be established, such as the centre line on straight section of road, long mitre drain outlets, establishment of structure line, etc.

*Procedure:*

- Fix ranging rods at the two ends of the straight (make sure that the ranging rods are fixed vertically).
- While sighting from one end to the other let an assistant move the third ranging rod at any distance between the two end rods until all three rods are in one straight line. Then place a peg to fix the point.

For example, to establish the centre line for a road along a straight section you must fix centre pegs at intervals of 10 metres; the picture below shows you the procedure.

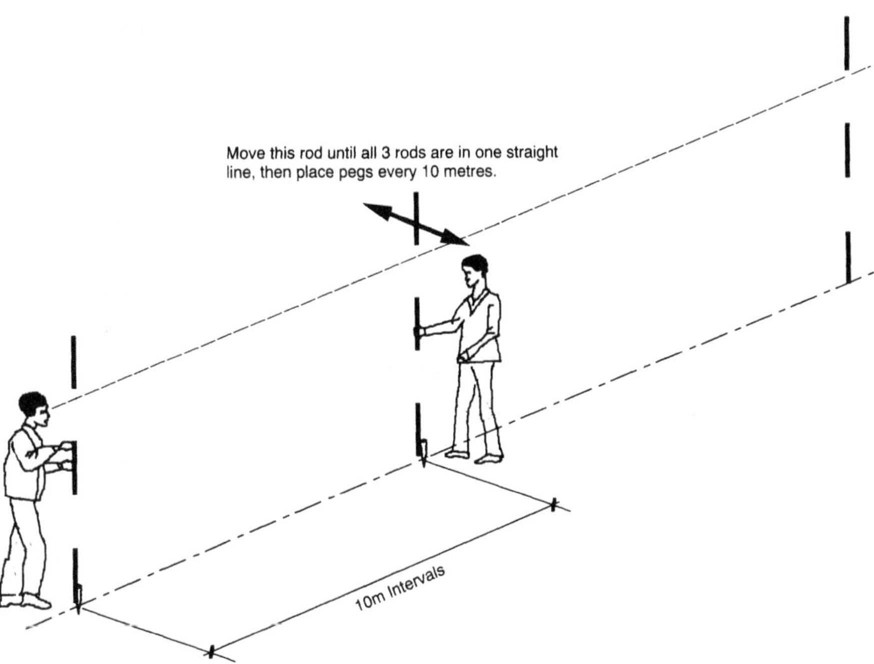


Figure 5.8 *Setting out centre line*

*Control of vertical alignment*

In order to achieve a reasonably smooth and aligned road surface without unnecessary ups and downs it is necessary to control the levels. In some labour-based projects the so-called 'slotting method' is used, while in other projects the 'profile board' method is used for this purpose. However, the principle measuring technique is the same, whether the boning rod or the profile board method is used. For you as a regravelling or maintenance contractor this technique will be important for the following reasons:

- to check the levels of the carriageway before and during reshaping of the road (no bumps, no dips)
- to check the uniformity of drain bottoms (are there any dips?)
- to check culvert inlets, outlets and end of outlet drains (same gradient).

*Procedure:*

- Fix boning rods/profile boards at the two ends of the straight you want to check, assuming that those two points have the level you need to maintain. If you use profile boards make sure that the two end boards are fixed at the identical measure on the rod, e.g. 1.30m from the ground.
- While sighting from one end to the other, let an assistant place the third boning rod/profile board at any point you want to check between the two end rods, e.g. the first rod is at the culvert inlet, the second at the end of the outlet drain and third is used to check whether the outlet of the culvert between is in line with the two end points, which would indicate that the entire culvert drain has the same gradient.

Figure 5.9 shows the procedure using boning rods; the same principle applies with profile boards.

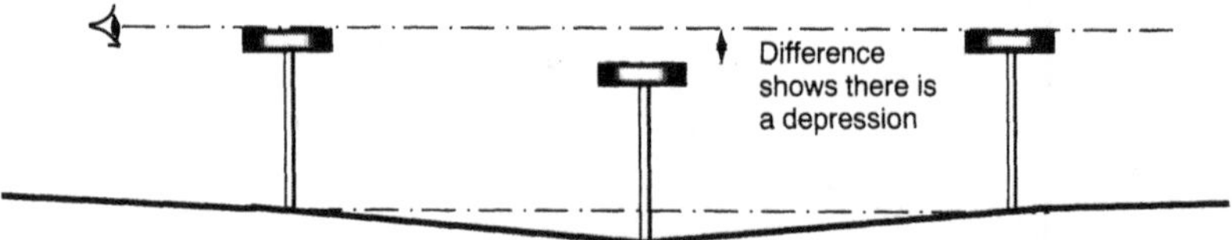


Figure 5.9 *Using boning rods*

*Checking or setting gradients*

For regravelling as a periodic maintenance operation and for routine maintenance you as a contractor generally do not expect to have to construct new road sections or drains. However you will frequently have to check gradients, especially of mitre drains and culvert outlets. In cases where they are not in accordance with the standards you will probably have to re-establish the correct gradients. More often than not drains silt up because the gradient is less than 2%, or the bottom of the drain has not got a uniform gradient (dips). If a drain has often to be de-silted it can become very costly for routine maintenance. It would probably be more cost-effective if, at the beginning of the contract, the gradients of all doubtful drains were checked and re-established if necessary. Sometimes cleaning or re-establishment of drains is included as part of the contract, and it will then be important to know how to check or set gradients using simple methods.

*Procedure for finding the gradient of a slope, e.g. a culvert outlet:*

- Fix ranging rods vertically firmly into the ground at the two end points of the slope.
- Tie the string line at the 1m mark of the ranging rod at the higher point of the slope.
- Fix the string line to the lower ranging rod, hook the line level at the middle point between the two ranging rods and move the string line at the lower point ranging rod up or down until the level bubble is exactly in the middle. Mark this level at the lower ranging rod, turn the line level around and mark the level again. Measure the middle of the difference of the two marks; this is the exact horizontal level transferred from the higher to the lower ranging rod.
- Now measure the difference between your horizontal level mark and the one metre mark at the ranging rod ( = D).
- Measure the exact distance (length) between the two ranging rods ( = L).
- Calculate the percentage of the slope; the distance between the two ranging rods represents 100%. The calculation is therefore as follows:

  100% divided by L multiplied by D = the percentage of the slope. Use centimetres for all measurements.

Figure 5.10 shows you the principle of the procedure using string and line level.

In case you have now checked the existing gradient of a culvert outlet using the above described method and the result shows that your outlet has a gradient of only 1.5% instead of the required 3%, you need to re-establish the 3% gradient. You therefore have to calculate the required difference D to achieve the required 3%:

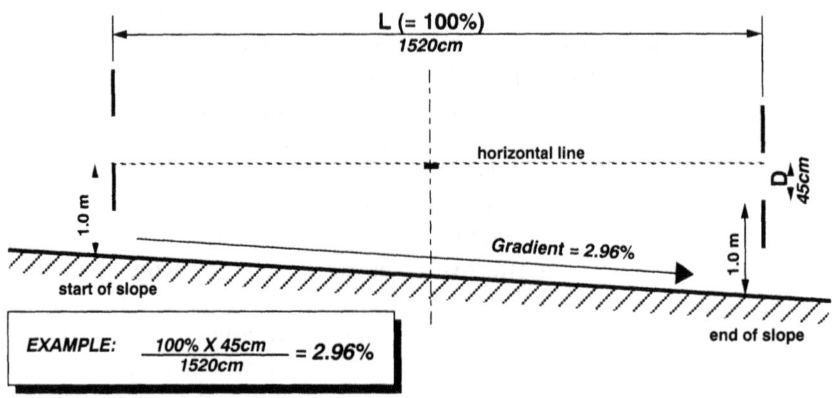


Figure 5.10 *Finding a gradient*

- The distance L is still the same and represents 100%.
- Calculate D : = divide L by 100% and multiply by 3%.
- Fix the string level with the line level to the lower ranging rod so that the line is horizontal, and mark the point on the ranging rod.
- Now add D to 1m and measure from the level-mark downwards. You will see that in order to be able to measure this new height you need to dig a small slot next to the ranging rod. Dig the slot in small steps until you can measure the exact height (D + 1m). The bottom of this slot is now at the required level.
- In order to transfer the gradient uniformly to the entire drain bottom you have to use boning rods or profile boards applying the method you have learned above, i.e. set a boning rod at each end point, every few metres dig a small slot, set the boning rod at the bottom of the slot and deepen the slot until all three boning rods are in line with one another.

Labourers could now follow these slots when excavating the ditch. The site supervisor would have to check the uniformity of the ditch bottom from time to time using the boning rods.

Figure 5.11 shows the procedure, including the calculation of a practical example.

*Setting-out horizontal curves*

There are four basic methods of setting-out curves that may be used for labour-based road construction. They all require the use of simple measuring aids, such as ranging rods, tape measure, strings and pegs. These methods are summarized in Table 5.3. (page 88)

These methods are used for construction or heavy rehabilitation of roads where the alignment needs to be fully established.

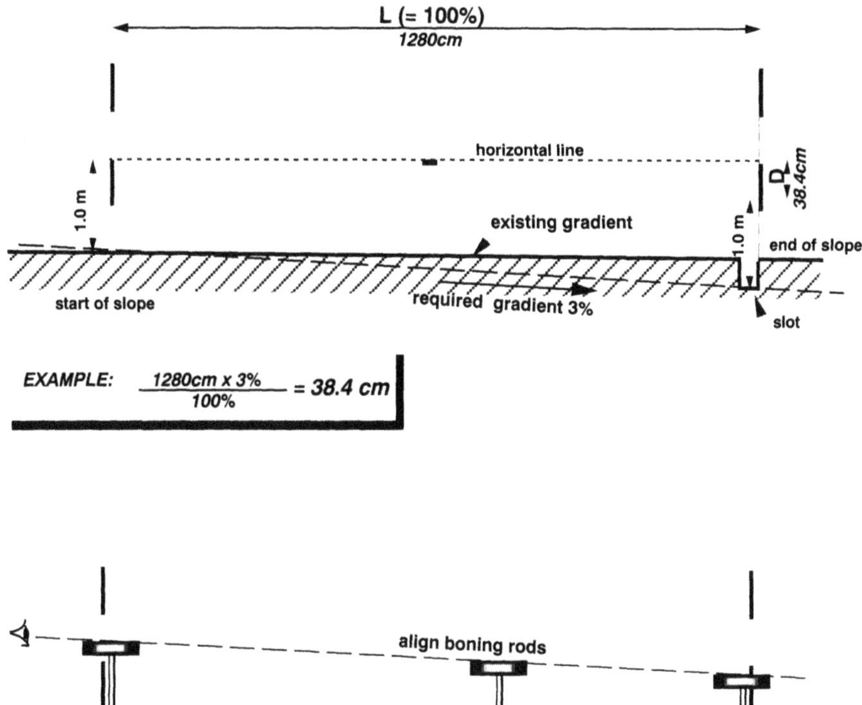


Figure 5.11 *Establishing a gradient*

It will not be necessary for you to know all the details concerning these methods, as the alignment of the roads you have to work on are already established.

If you are involved in regravelling works it is sufficient, when reshaping a road, to follow the existing alignment of curves with minor corrections only. As a contractor you need to have a method which allows you quickly to check the alignment of curves, to establish the centre line in curves and, if necessary, to correct or re-align slightly.

*Procedure:*
The setting-out procedure for curves is simple and straightforward. First the beginning and the end of the curve have to be

**Table 5.3** Curve setting-out methods

| *Method* | *Application* |
|---|---|
| String method | Suitable only for curves with a radius of less than 30m. The area must be flat and free from obstructions. Most suitable for junction curves and hairpin bends. |
| Quarter method | Suitable for short curves where a stringline can be stretched unobstructed between the ends of the two straights. |
| Tangent method | Suitable for any curve where the alignment turns through less than 90°. The intersection point for the two straights and the area between it and the road must be flat and free of obstructions. |
| Offset method | Suitable for any curve. However, it requires a trial-and-error approach when applied to an existing alignment. |

located and a centre peg established at these ends. Then approximate centre pegs are set in between at intervals of 5 metres; the existing ditch or edge of camber can be used as a guide to measure from. A string is attached and laid along all these pegs. All curve segments have then to be checked by eye whether they look similar. Where abnormal deviations can be found the nearest centre peg has to be re-set until the entire curve looks uniform.

Figure 5.12 explains the procedure in detail.

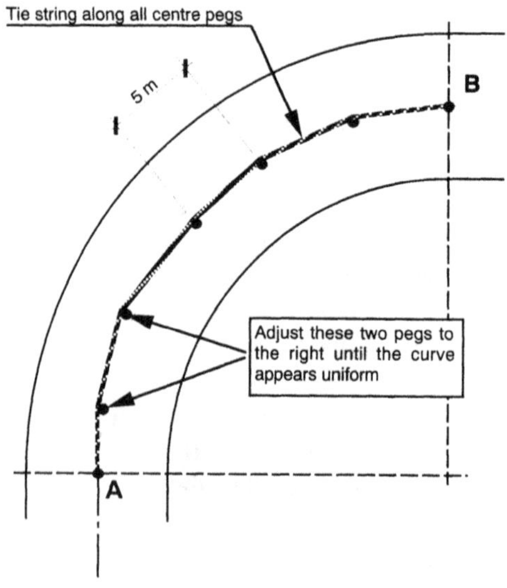


Figure 5.12 *Curve setting-out methods*

1. Establish the centre-line pegs at each end of the curve (A and B).
2. Set intermediate pegs at 5 metre intervals along the approximate centre-line.
3. Tie a string along all centre pegs.
4. Adjust pegs which do not appear to be on a smooth curved line until the entire curve appears to be uniform.

# Drainage and Erosion Control

On earth and gravel roads an effective drainage system is most important.

Labour-based methods are especially suitable for drainage work. Accuracy and attention to detail can be achieved better by labour-based methods than with heavy equipment.

Proper attention has to be paid to setting out gradients, controlling levels and shaping of drains. Standards for drains are usually provided by the project clients and we suggest that you compare them with the principle drainage features which we have already discussed in Chapter 1. Specific maintenance drainage aspects are dealt with further in Chapter 7 – Routine Maintenance, and Chapter 8 – Regravelling.

In this section we will discuss some important aspects which are essential to know when it comes to regravelling and routine maintenance contracts.

## DRAINAGE PRINCIPLES

Various drainage measures are necessary to deal satisfactorily with rainwater falling on or near the road. Rainwater is the main cause of damage to earth and gravel roads; however, a good drainage system will reduce its effects significantly and minimize future road maintenance requirements. Water can damage the road in two principal ways:

- by weakening the soil (reducing the traffic-bearing capacity)
- by erosion and silting (damaging the road and reducing the effectiveness of the drainage system).

The drainage system must therefore collect all rainwater and dispose of it quickly in a controlled manner to minimize road damage. It should enable the road to dry out quickly after rain and regain its full strength.

The drainage system will normally require most of the following components to function effectively. It is important to make sure that they are provided at the necessary locations.

| | |
|---|---|
| Road Surface Camber: sheds water from the road surface (see Chapter 1) | Figure 5.13 |
| Side Drains: collect water from the road surface and adjoining land (see Chapter 1) | Figure 5.14 |
| Mitre Drains: lead the water out of the side drains and safely disperse it on to adjoining land (see Chapter 1) | General layout of mitre drains see Chapter 1, Figure 1.2.<br>For construction details see explanations below. |
| Catch Water Drains: catch or intercept surface water flowing towards the road from adjacent land, and lead it away (see Chapter 1) | Figure 5.15 |
| Culverts (Drifts/Splashes): allow water to pass from one side of the road to the other (see Chapter 1) | Figure 5.16 |
| Scour Checks: prevent erosion in side drains by slowing down the water. (For construction details see below) | Figure 5.17 |

*Mitre Drains*

Mitre drains (or turn out drains) lead the water away from the side drains to the adjoining land and this must be achieved so as to avoid causing erosion at the discharge point. Therefore mitre drains should be provided as frequently as possible so that the accumulated water volume in each drain is not high. Some countries have worked out exact guidelines about the different intervals on certain gradients. However, as a principal guideline it is recommended that mitre drains are constructed at intervals of 20 metres along the road alignment ('little and often' should be the guiding principle for water discharge).

In highly populated and agricultural areas the discharge water should be channelled to field boundaries where possible, in order to avoid damage to farm land. The minimum width of the mitre drains should be 40cm and they should have a gradient between 2% and 4%. Gradients should be carefully checked to ensure that they drain positively within these limits.

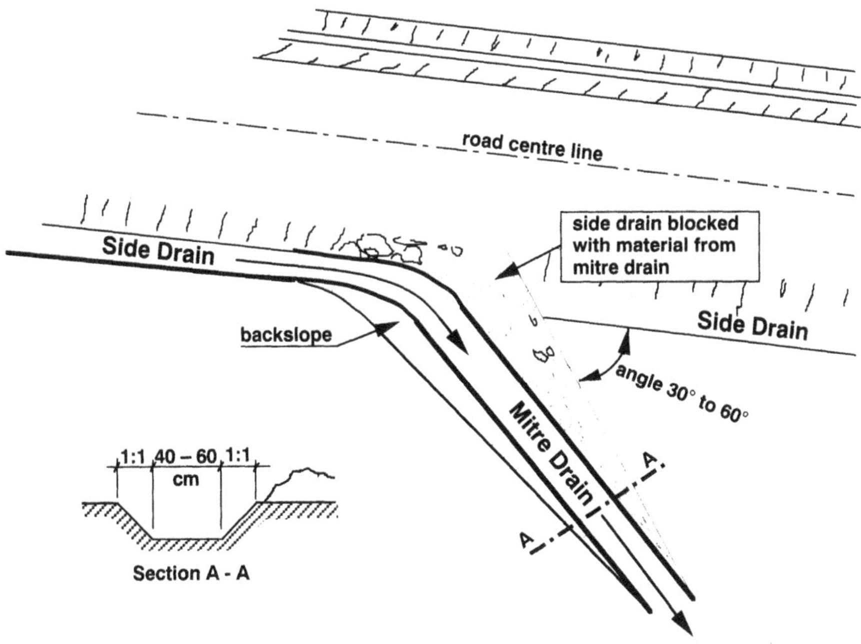


Figure 5.18 *Mitre drain*

If the surrounding land is flat and only allows gradients below 2%, the outlets must be set out very carefully. It is important to find areas where the water can discharge without forming ponds in the drains. The width of the drains must be increased and should be at least 1 metre wide.

Some of the excavated soil is used to block the downhill side of the side drain to ensure that the water flows into the mitre drain (see Figure 5.18 above).

*Scour Checks*

Where drain gradients are steeper than about 4% the water flows at high speed. Therefore, if no protective measures are taken scouring is likely to occur on erodible soils. The simplest way of dealing with scouring is by reducing the volume of water (with mitre drains at frequent intervals). In addition, scour checks can be constructed to reduce the velocity of the water. They hold back the silt carried by the flow of water and provide a series of stretches with gentle gradients interrupted by small 'waterfalls'.

Scour checks are not usually provided on roads which are constructed using equipment-based methods. During maintenance, motor graders cleaning the drains with the blade would destroy the scour checks. Therefore roads which are constructed using labour-based methods should also be maintained by labour.

Scour checks are constructed either with natural stones or with wooden pegs. The level of the scour check must be a minimum of 20cm below the edge of the carriageway in order to avoid the water-flow being diverted out of the side drains (see Figure 5.19). The constructed scour checks have therefore to be controlled with a template (which can be part of the ditch-slope template (see Figure 5.20). The intervals at which scour checks are constructed depend on the gradient, as shown in Table 5.4. Determine the gradient using a string and line level. Scour checks should not be constructed on roads with gradients of less than 4%. This will encourage too much silting of the side drains and lead to road damage.

**Table 5.4** Scour check spacing

| *Gradient of road* | *Scour check spacing* |
|---|---|
| 4% or less | not required |
| 5% | 20m |
| 8% | 10m |
| 10% | 5m |

After the basic scour check has been constructed an apron should be built immediately downstream using either stones or grass turf pinned to the ditch invert with wooden pegs. The apron will help resist the forces of the waterfall. Sods of grass should be placed against the upstream face of the scour check to prevent water seeping through the scour check and to encourage silting behind the scour check. The long-term objective is to establish a

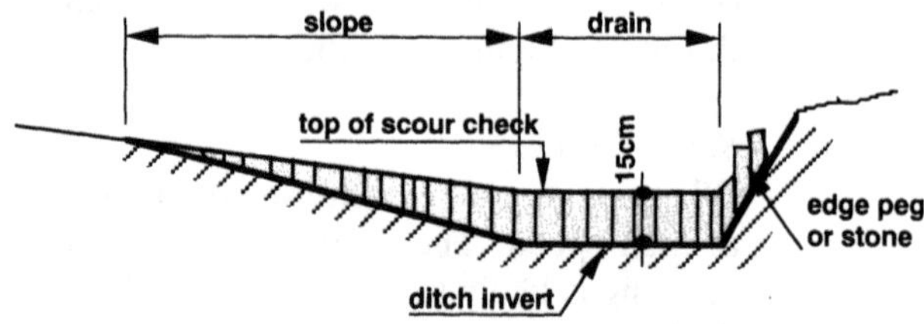


Figure 5.19 *Scour check profile*

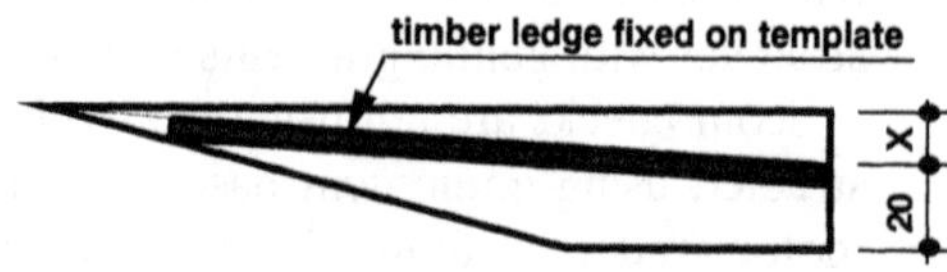


Figure 5.20 *Scour check template*

complete grass covering over the silted scour checks to stabilize them.

Figure 5.21 sets out the detailed working steps.

*Procedure for scour check construction:*

1. Identify road sections where ditch gradient is more than 4% using a line-level; in such cases construct scour checks.
2. Identify the exact ditch gradient and space scour checks according to gradient (see Figure 5.21).
3. Cut pegs (minimum 50cm long) and/or prepare stones.
4. Construct scour check with the correct profile. Use the scour check template for control.
5. Construct stone apron below scour check of minimum 40cm length. Dig stones into the ground.

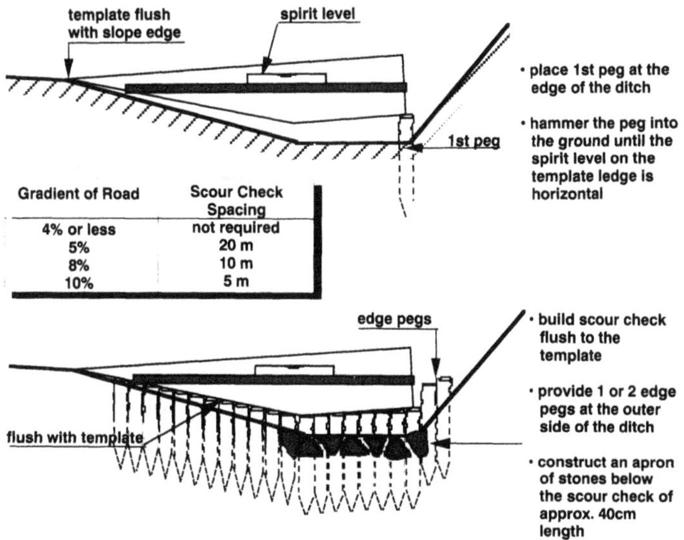


Figure 5.21 *Construction of scour check*

# Productivity

Productivity is the measure of the amount of work that can be done in a specified period of time. For example, 1 labourer excavates $4m^3$ of soil in 1 day, which is the same as $4m^3$/work-day. For pricing and bidding in the construction business, productivity rates are also calculated as hourly rates; this allows a more accurate calculation. In our case we would assume that a working day would have 8 hours and the productivity rate would therefore be $0.5m^3$/hour.

Work on site cannot be planned and controlled unless productivity rates for all activities are known. Only when these rates are defined can you set targets, plan the size of operation gangs, and have a measure to control the quantity of work achieved in a specified period of time. This is particularly important for you as a contractor. In a labour-based project, where labour is the main means of production, the achieved productivity makes all the difference to whether you make a profit or not.

*Example:*
*You have been given a contract for road construction. In your tender you have quoted the price for excavation of soil to form the road base as 0.5 NU/m*$^3$*, assuming the soil is soft and easy to excavate. You had calculated the price on the assumption that the daily output of a labourer would be 4m*$^3$ *and the daily wage-rate of the labourer, including overheads, would be 1.75 NU. The total excavation was estimated to be 5600m*$^3$ *and therefore the total sum in your contract for this activity was 2800 NU. You had expected to make a profit of 350 NU. However, the soil is much harder than you assumed, and the daily productivity was only 3m*$^3$ *per labourer. Instead of making a profit of 350 NU you will now make a loss of 565 NU as the labour costs plus overheads amount to 3265 NU. If you deduct the expected profit of 350 NU you are still getting a net loss of 215 NU which you have to pay from your own pocket.*

In order to achieve the planned productivity it is necessary to motivate and encourage the labourers by giving incentives. In most projects this is achieved through the payment system.

For the management of casual workers, specific incentives have to be provided in order to achieve a high and even level of productivity. Three basic incentive schemes exist for labour payment in road construction and maintenance:

*Task work system*
A fixed daily wage is given in return for a fixed quantity of work. The worker is free to go home as soon as the given quantity of work has been done and approved. Tasks can be given to individuals or to groups.

*Piece work*
The worker is paid a fixed sum per unit of output (e.g. 100 NU per cubic metre of excavation). The daily output is left to the discretion of the worker, who can thus increase his daily earnings by producing more. This system requires good control and flexible payment conditions.

*Daily paid work*
The worker is paid a fixed sum of money for a day's work. It is left to the supervisor to utilize the worker as efficiently as

possible. This system is sometimes used in labour-based projects for activities which are not easily quantifiable (e.g. boulder removal) or where the mere presence of the worker is required (e.g. watchmen).

The most successful incentives for different programmes has been the organization of work in the task-work system. The Table 5.5 shows a list with average production targets (task rates) for possible activities of labour-based road construction and gravelling. This list should be used as a guideline only, and needs to be adapted to the specific local conditions.

**Table 5.5** Average productivity targets for labour-based earth road construction

| *Activity* | *Task rate* | *Remarks* | *Tools* |
|---|---|---|---|
| Bush clearing | 50–300m²/wd | quantity according to the nature of the bush (dense, medium, light) and the experience of the site supervisor | bush knife, axe |
| Grubbing | 100–200m²/wd | according to density | slasher, hoe, shovel |
| Tree and stump removal | according to experience | for small stumps in m², for big stumps in No. | axe, saw, mattock, pickaxe, shovel, rope |
| Boulder removal | according to experience | | sledge-hammer, crowbar, pickaxe, rope, shovel, chisel/ wedges, plugs and feathers |
| Slotting | 4–8 slots/wd | according to terrain | pickaxe, shovel, straight edge, spirit-level |
| Excavation<br>○ loose, not sticky<br>○ firm soil<br>○ hard or stony soil | <br>4–6m³/wd<br>3–5m³/wd<br>2–3m³/wd | <br>throwing distance 0–4m<br>throwing distance 0–4m<br>throwing distance 0–4m | <br>hoe, shovel<br>forked hoe, hoe, shovel<br>pickaxe, crowbar, pickaxe, shovel |
| Loading<br>○ loose soil on to the wheel-barrows<br>○ loose gravel on to the trailer/ cart | <br>10–15m³/wd<br>7–10m³/wd | | <br>shovel<br>shovel |

(*continued over*)

**Table 5.5** *(cont.)*

| *Activity* | *Task rate* | *Remarks* | *Tools* |
|---|---|---|---|
| Ditching, sloping<br>○ loose, not sticky<br>○ firm soil<br>○ hard or stony soil | <br>4–6m³/wd<br>3–5m³/wd<br>2–3m³/wd | excavation and throwing the soil on to the centre line | hoe, shovel<br>forked hoe, hoe, shovel<br>pickaxe, crowbar, shovel (+ template) |
| Backsloping | 15–25m/wd | depending on height of cut and soil type/excavation and throwing the soil on to the centre line | hoe, pickaxe, shovel |
| Spreading<br>○ loose soil<br>○ loose gravel | <br>12–18m³/wd<br>9–12m³/wd | | shovel, hoe, spreader or rake<br>forked hoe, hoe, shovel, spreader or rake |
| Camber formation<br>○ loose soil | <br>10–12m³/wd | soil form ditching heaped on the centre line | hoe, shovel, spreader or rake, camber board and spirit-level |
| Scour checks | 4–8No/wd | including preparation of stones or pegs | hoe, pickaxe, hammer, template |
| Collecting stones | 2–3m³/wd | within a radius of 50m | wheelbarrows |
| Culvert laying | 15/line | depends on the number of culvert rings and the type. Group task for excavation, laying rings, back-filling, compaction and headwalls | pickaxe, hoe, shovel, ropes, earth rammer, boning rods, culvert bed template |
| Hauling distance:<br>10–30m<br>50m<br>75m<br>100m | <br>10–12m³/wd<br>7.5–9m³/wd<br>5.5–6.5m³/wd<br>4.5–5.5m³/wd | | <br>wheelbarrows<br>wheelbarrows<br>wheelbarrows<br>wheelbarrows |
| Erect cement masonry | 0.5m³/wd | including preparation of stones and mortar | |
| Erect dry masonry | 1m³/wd | including preparation of stones | |
| Mix and place concrete | 0.5–1m³/wd | | |

## Low-cost Structures

Using labour-based methods also means using appropriate technology for the construction of structures. It is the aim of most labour-based programmes to construct low-cost structures, wherever possible utilizing locally available material, such as stones, sand and timber.

The location and type of structures should be adequately set out at the beginning of the planning stage. Structures usually account for a high proportion of the cost of a road and is therefore important when estimating the total resources required for a particular road.

For a maintenance contractor it is important to know the various types of low-cost structures, to understand their function, how to inspect them for failures and how to repair them. Some of the principal low-cost structures were introduced in Chapter 1, while inspection and repair will be discussed in Chapters 6 and 7.

## Gravelling

In terms of site management, gravelling is the most demanding operation in labour-based projects. The mixture of large labour forces together with several units of equipment is a complex management task, as labourers depend on the equipment to perform efficiently and on the other hand equipment utilization depends on the efficiency of labourers and operators.

Regravelling is very similar to the initial gravelling operation, and we will deal with this in detail in Chapter 8. This section therefore presents only an outline to give you as a contractor a first idea about the complexity of the gravelling operation.

As with the initial construction work, the gravelling activities have to follow each other in logical sequence, and remain independent of the choice of equipment. Whether wheelbarrows, lorries, donkey-carts, oxen-carts or tractor/trailer combinations are used to haul gravel, the basic activities and the work organization remain the same. The maximum utilization of the available equipment plus labour force must be the goal of good site management.

Gravelling consists of the following major activities:

- site installation
- preparation of quarry access road
- opening of quarry
- excavation and stockpiling of gravel
- reshaping/rehabilitation of road to be gravelled
- gravel loading
- gravel hauling
- gravel off-loading and spreading
- compaction
- gravel stockpiling for maintenance purpose

The improvement of the quarry access road, the quarry and road preparation must be carried out before the equipment is brought to site so that the actual gravelling work can then start immediately. Good timing is therefore required.

The size of the different labour gangs depends mainly on the amount of gravel to be hauled in a day. This, in turn, depends mainly on the available hauling equipment and the hauling distance.

Gravelling requires a careful daily work plan giving information on:

- required labour
- required equipment
- average hauling distance
- quantity of gravel to be hauled
- length of road to be gravelled
- necessary preparatory work

The quantity of inputs required depends to a great extent on three factors:

- available hauling equipment
- the average hauling distance
- task rates for workers

The output depends on:

- number of trips per equipment and day
- quantity of material transported per trip
- average speed of haulage equipment

Once this data is established, various types of planning tables can be prepared. Such tables contain the following information:

- hauling distance
- trips per item of equipment per day
- average number of equipment in operation
- planned daily trips in total
- metres gravelled per day
- number of labourers required

# Alternative Pavements

In certain cases it may not be possible to find gravel of satisfactory quality close to the road, and it would be uneconomical to haul it from far away. Other roads may have difficult sections, such as steep hills or pockets with poor soils. Certain projects may also decide to use a different surface material (dust- and clay-free) rather than gravel when a road passes through a village or town. Alternative pavements which can also be constructed using labour-based methods may be used in such cases.

The principal alternatives are:

- hand-packed stones
- concrete running strips
- concrete block paving
- surface dressing

If these alternatives are applied over short sections the overall cost of the road may not be significantly increased, but maintenance requirements may be greatly reduced in the long run.

These alternative surfacing methods require additional supervision and careful planning to achieve an adequate construction quality.

# Planning and Reporting Systems

To ensure proper planning and monitoring of all field activities a clearly structured and transparent management and reporting system is required.

Whatever system of payment is applied (daily-paid, task-work, piece-work) the management and reporting system has to allow for:

- planning targets on productivity
- planning targets on resource inputs (labour, material, equipment)
- planning technical working details
- recording productivity and progress actually achieved
- recording used resources
- recording costs
- recording labour performance actually achieved
- recording site overheads

The system must allow for this detailed planning and reporting on all activities undertaken. Further, the system should be designed in such a way as to filter only the important data for further transmission to the next higher level. Such summary reports should provide the supervisors with enough information to allow a thorough desk-monitoring. The site plans and reports are the basis for controlling production. The system must be detailed enough to enable the site supervisors to plan and record all activities on a daily basis, and must provide sufficient information for the contractor to be able to control and monitor the progress on site. Figure 5.22 may assist in visualizing the structure of a planning and reporting system.

The system should cover all important aspects which are necessary to plan and monitor quality, quantity and cost of all work elements. The following elements should appear in a comprehensive site planning and reporting system:

- labour attendance
- earth-work calculations
- overall targets and achievement
- activity planning and recording
- support (overhead) planning and recording
- productivity
- quality performance
- material consumption
- availability and utilization of site equipment

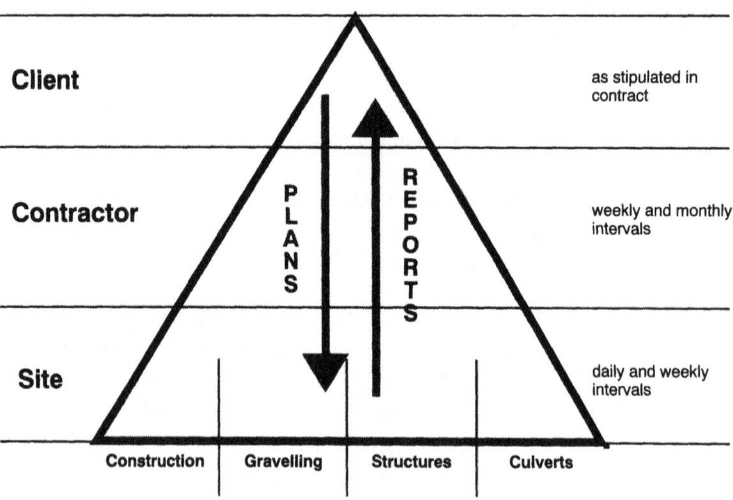


Figure 5.22 *Planning/reporting flow*

The forms should be simple to fill in and read in order to ease the work of the site supervisor.

We shall discuss relevant regravelling and routine maintenance planning and reporting systems in detail in the respective Chapters 7 and 8.

# Example: Construction activity steps

## RURAL ACCESS AND MINOR ROADS PROGRAMME, KENYA

Labour-based Road Construction in Kenya

Extract from: *Engineer's Orientation Course Manual, MRP*

Published by Ministry of Public Works, Department of Staff Training, Kisii Training School and Swiss Development Cooperation.

Edited and compiled by A. Beusch

Drawings by W. Illi

## BACKGROUND

The Kenya Rural Access and Minor Roads Programme is one of the most successful labour-based programmes in Africa. It started as early as 1974 and has since constructed and maintained some 12 000km of rural and minor roads. The programme has provided work to approximately 20 000 labourers each year and the quality achieved is considered to be adequate for this kind of roads.

*The main objectives of the project are:*

- o to improve access between farming areas and marketing centres
- o to expand the local road system at low cost
- o to provide meaningful employment opportunities to the people in whose areas the programme is implemented.

All construction activities are carried out by labour except the hauling of gravel, where tractor/trailer combinations are used. The labourers are set on equal task rates. The construction activities and their spacing and organization are uniform throughout the programme, which allows for effective training, planning and monitoring.

In order to be able to work with employees who are employed on a temporary basis (casual labourers) and to achieve uniform standards in terms of quality and quantity it was necessary to break down the activities into small and easily-controllable construction steps.

On the following pages the different work sequences with their activities are illustrated and described to detail, and are meant to show how a large-scale labour-based road construction programme can be easily planned and controlled.

*1. Setting Out*

The first activity is setting out the road alignment. This can be done with simple methods. Careful setting-out is important in order to produce a high quality road, and at the same time find an alignment where as little resources as possible have to be spent.

**Setting out of straight lines**
**Straight pegs, which should be painted in a bright colour for better visibility, are set in a row at a chosen distance. Ranging rods must always be set vertical.**
**Example: a) Placing of pegs for cross-section.**
**Extension of a straight line.**

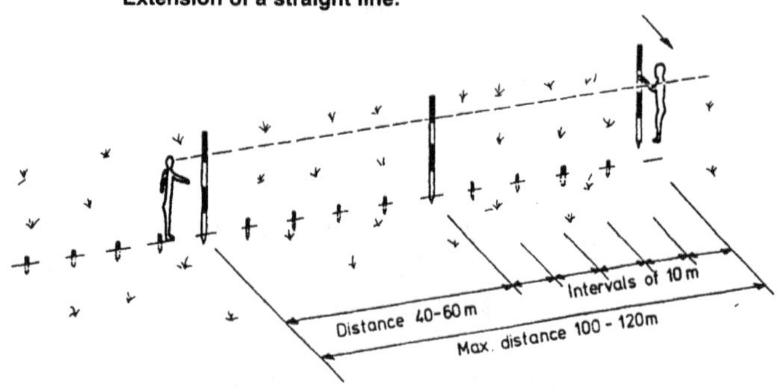


Figure A *Setting out*

*2. Bush Clearing*

This activity is usually carried out over the width of the road plus one or two metres on either side. Bush clearing consists of cutting and removing all bushes and shrubs within this area and disposing of the cuttings outside of the cleared area.

*3. Stripping and Grubbing*

This activity includes the removal of all grass, upper grass roots and other vegetation remaining after bush clearing (except trees) over the width that the earthworks will take place. All topsoil is also removed from the grubbed width, as this would weaken the road if included in its construction.

## BUSH CLEARING

13.00m

6.50m

6.50m

Chainage peg

New roadformation

Chainage peg (shows extent of bush clearing)

Centre peg

Existing road

Chainage peg:

Metres

Kilometre

Excavation to level 9.00m

Disposal of shrubs :

Outside the 13.00m wide strip of bush clearing, preferably on the lower road side.

Figure B *Bush clearing*

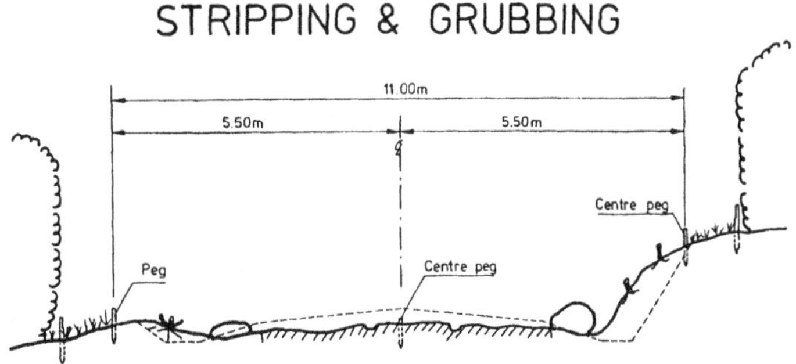


Figure C *Stripping and grubbing*

*4. Tree and Stump Removal*

This activity includes the felling of trees within the area previously cleared of bushes. The stumps must also be up-rooted and all material disposed of outside of the cleared width.

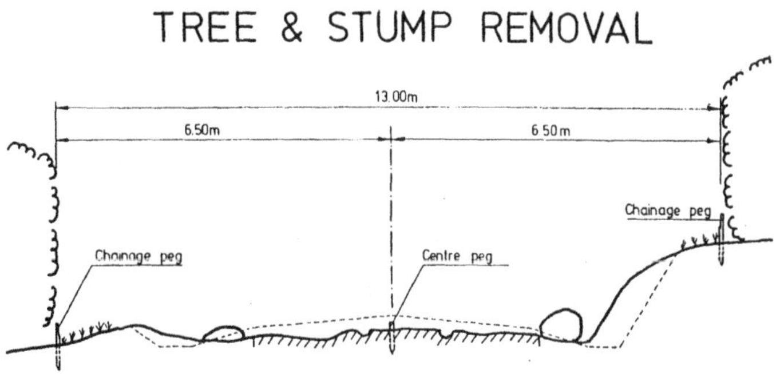


Figure D *Stump removal*

*5. Boulder Removal*

Boulders and rocks are encountered on some roads. Several methods can be used to deal with these: modify the alignment, remove boulders from road, bury boulders, split boulders using fire and water, split boulders using hand tools or split boulders using explosives.

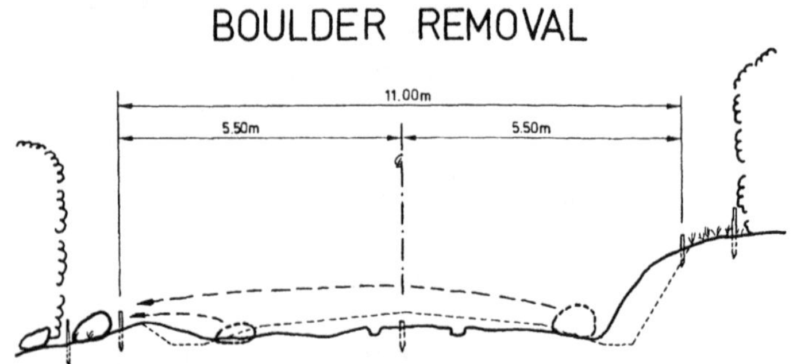


Figure E *Boulder removal*

*6. Slotting*

The slotting method balances the earth material on each 10 metre section of the road. It achieves this by taking a representative 'slot' across the road. The soil is redistributed and compacted within the 0.5m slot until the surface is level across the width of the road. These slots are then used as guides to:

- calculate the earthworks excavation quantities between each slot
- act as visual guide for the 'Excavation to Level' activity.

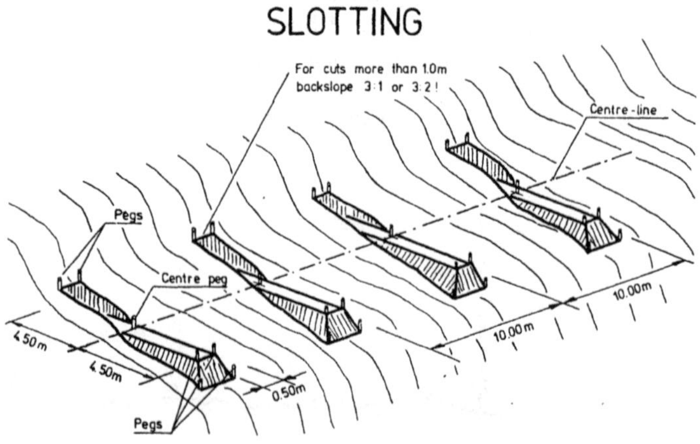


Figure F *Slotting*

*7. Longitudinal Balancing*

If the existing alignment is uneven the slots may also vary considerably in alignment along the road. To overcome this longitudinal balancing is carried out as a follow-up operation to the initial slotting. If three sequential slots are out of vertical alignment by more than 10cm (checked using boning rods), then the slots should be adjusted by transporting material between adjacent slots.

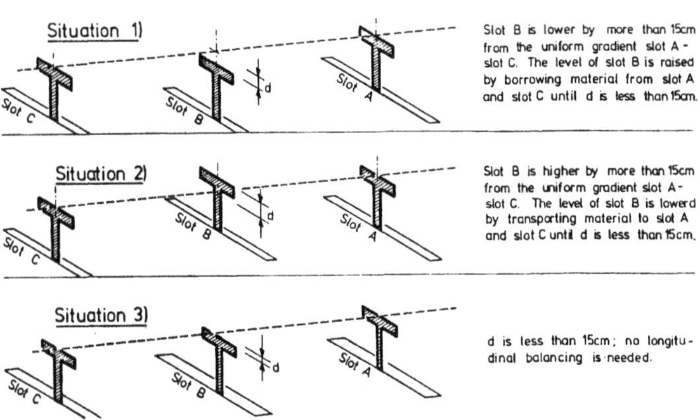


Figure G *Longitudinal balancing*

*8. Excavation to Level*

Once the alignment and slots are finalized then the bulk excavation to achieve the level terrace can be carried out. The slots act as a guideline for the excavation and filling of each 10m section of road to achieve a transversely-level terrace. The terrace is a platform on which to build up the road camber and drainage in an accurate and controlled manner.

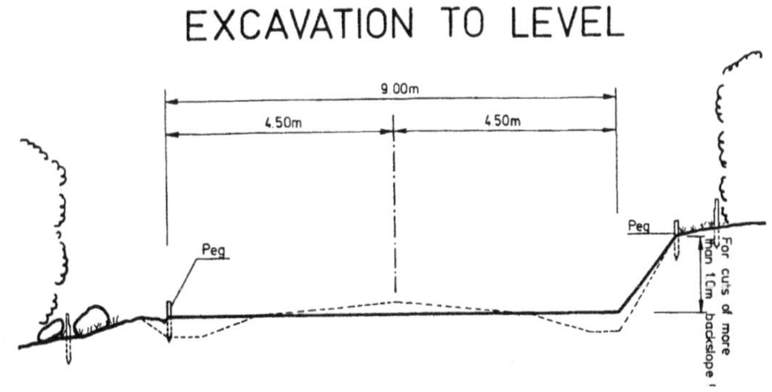


Figure H *Excavation to level*

*9. First Compaction*

If a roller is available the terrace fill should be compacted to avoid later consolidation, and to ensure good geometric quality for the following operations.

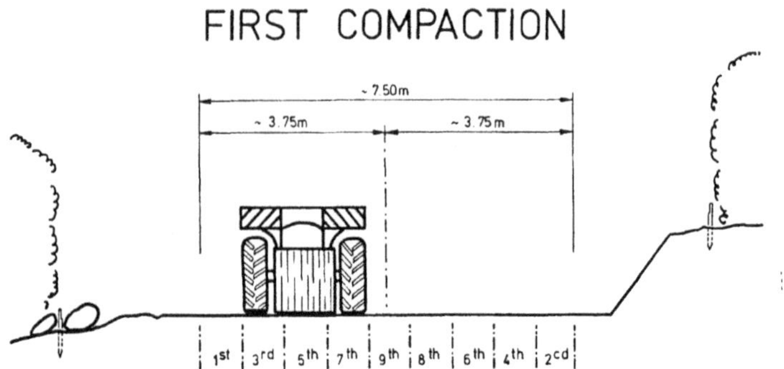


Figure I *Compaction*

*10. Ditching*

The side drain operations enable the side drainage system and the road camber to be constructed. The side drains are normally constructed in three simple excavation steps (ditching, sloping, and backsloping) to facilitate geometric and productivity control of the operations. The activities involved are:

- Ditching
- Spreading
- Second compaction
- Sloping
- Backsloping
- Camber Formation
- Final (third) compaction

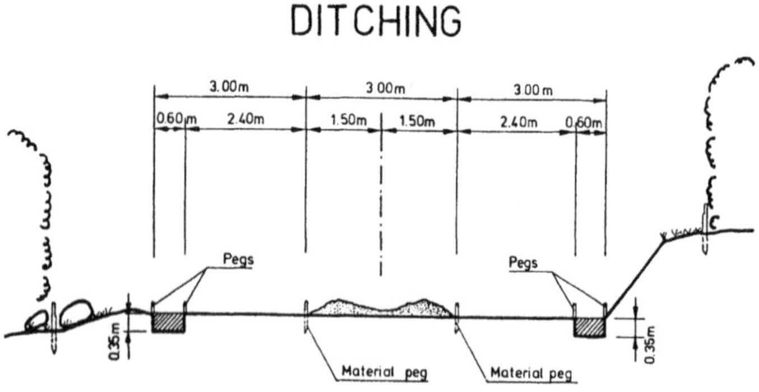


Figure J *Ditching*

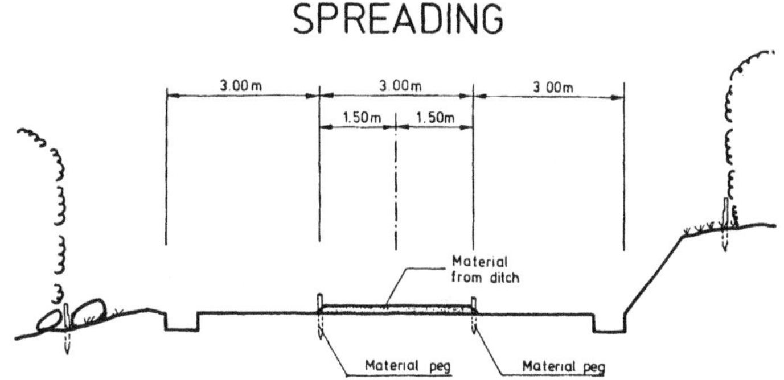


Figure K *Spreading*

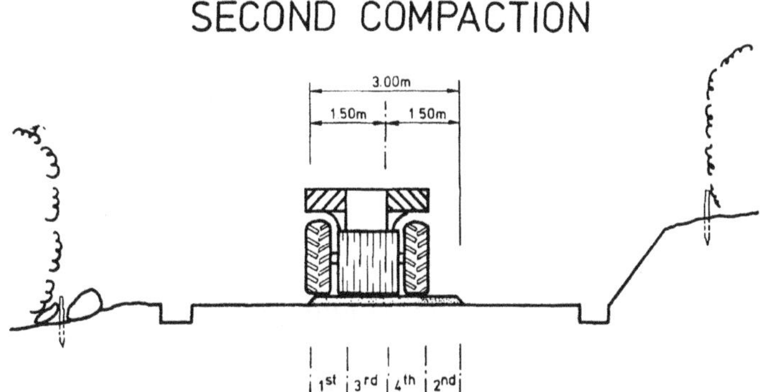


Figure L *Second compaction*

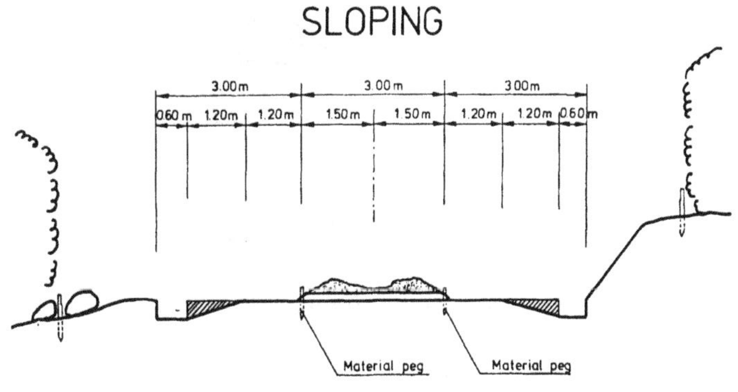


Figure M *Sloping*

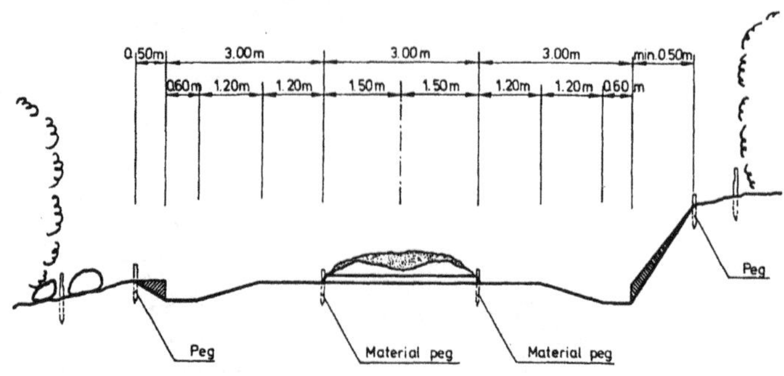


Figure N *Backsloping*

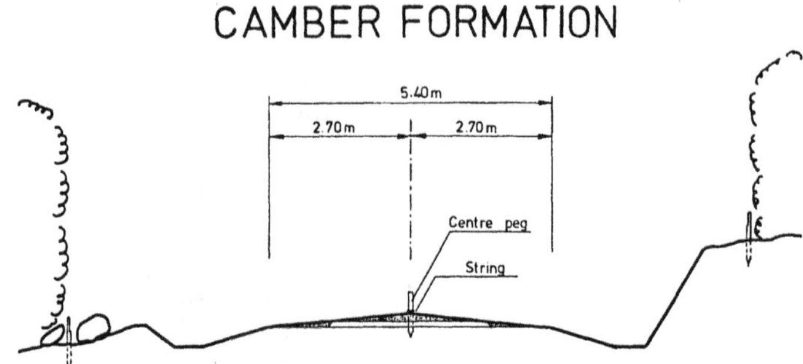


Figure O *Camber formation*

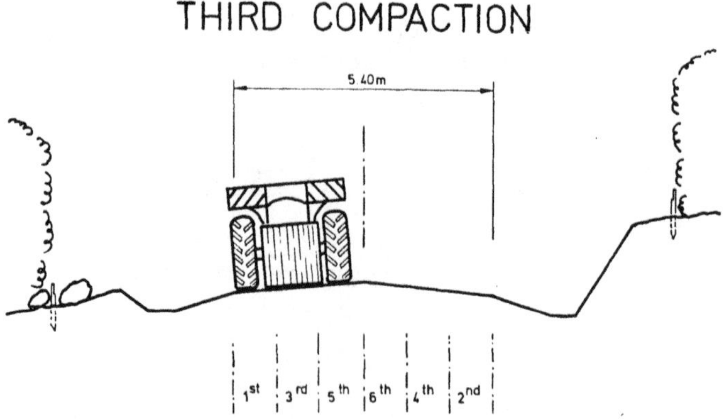


Figure P *Third compaction*

*11. Mitre Drains*

Mitre drains (or turn-out drains) lead the water away from the side drains to the adjoining land. This must be achieved in a manner to avoid causing erosion at the discharge point. Therefore mitre drains should be provided as often as possible so that the accumulated water volume in each drain is not too high.

MITRE DRAINS

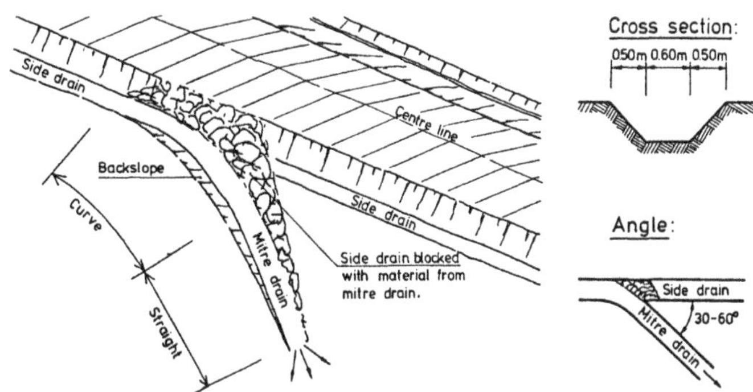


Figure Q *Mitre drains*

*12. Scour Checks*

Scour checks are usually constructed of natural stones or wooden stakes. The level of the scour check must be a minimum of 20cm below the edge of the carriageway in order to avoid the water flow being diverted out of the side drains.



Figure R *Scour checks*

*13. Culverts*

Pre-cast concrete culverts with a diameter of 60cm are placed at an average of every 200 metres. The exact site of the culvert has to be determined at each low point in the vertical alignment and at each location that a relief or crossover culvert is required.

## CULVERTS

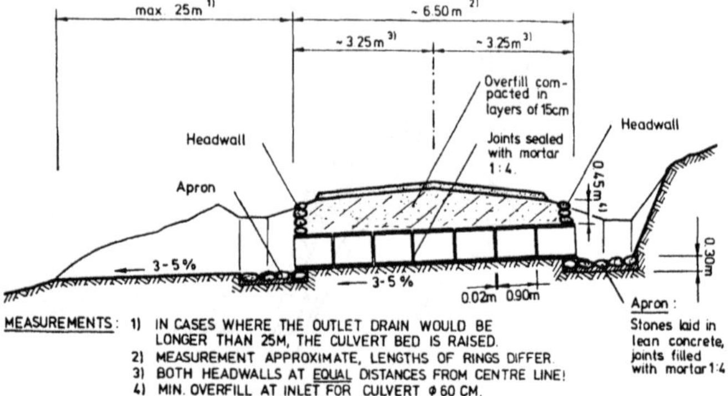


Figure S *Culverts*

Other activities which occasionally need to be carried out are excavation of catch water drains, planting of grass or turfing on slopes to prevent from erosion, lining of side drains, etc.

SECTION B

# PRACTICE

# CHAPTER 6: ROAD MAINTENANCE

## Learning Objectives

After you have completed this chapter you should know:

- the purpose of and need for road maintenance
- the basic causes of deterioration of gravel and earth roads
- the basic maintenance operations
- the priorities for road maintenance
- how to assess road maintenance requirements and plan for maintenance interventions

## Introduction

The three main purposes of road maintenance are:

- to prolong road life and postpone the day when renewal will be required
- to reduce the costs of operating vehicles on roads
- to keep roads open and enable greater regularity, punctuality and safety of road transport services.

Management of maintenance operations is a complex and demanding undertaking for responsible road authorities. You as a maintenance contractor can look forward to a good business future if you can perform well in organizing and implementing maintenance contracts. To do this, you must be fully aware of basic maintenance management aspects, and the reasons behind them; so you need to understand how roads deteriorate, before you can set up proper maintenance systems and establish an effective maintenance management process.

## Mechanisms of Deterioration of Roads

As when a car breaks down, there may be many reasons why a road deteriorates, and you must go through a number of logical steps to diagnose the problem. Once you know why a road, or part of the road, deteriorates, it will be much easier to apply the correct measures to repair it.

For example, suppose a culvert outlet always silts up after heavy rainfall. If you did not investigate the cause of this frequent silting up, or diagnosed it wrongly, you would have to allocate a labourer after every storm to clean out the drain. This is very costly and wasteful. But if you are a professional contractor who understands the principles of road drainage, you would probably carry out a few investigations and find out that this culvert outlet has a gradient of only 0.5%, although it would be possible to have a gradient of 3%. The correction of this gradient would initially cost the equivalent of 4 workerdays, but afterwards you would hardly ever have any cleaning work to do and you would save a substantial amount of working time which would result in an increased profit.

Let us first discuss what deterioration and failure of roads actually means.

## DETERIORATION OF ROADS

Deterioration of roads is the worsening of roads over a period of time due to a variety of causes. Deterioration leads to defects of the road structure.

The three main mechanisms for degradation are:

- degradation of carriageway (Rutting, Potholes)

Figure 6.1 *Degradation of carriageway*

- silting of the drainage system

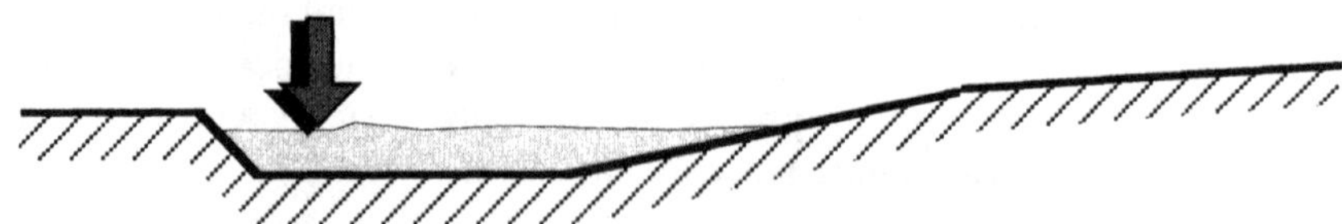

Figure 6.2 *Silting of the drainage system*

- erosion of the drainage system

Figure 6.3 *Erosion of the drainage system*

The main causes of deterioration are several, and can be summarized as follows:

- rainfall
- steep gradients
- flat gradients
- traffic
- pavement construction
- vegetation.

Not each factor affects the different parts of the road structure in the same way:

- Rutting, potholes and deformation of the carriageway is mainly caused by traffic.
- Loss of gravel is mainly caused by steep gradients and rainfall.
- Erosion of the drainage system is caused by rainfall, steep gradients and a lack of vegetation.
- Silting of the drainage system is also caused by rainfall, flat gradients, and to some extent, by vegetation.

The first two factors affect the carriageway, and the second two the drainage system. So it is possible in the planning of maintenance to separate carriageway maintenance from the maintenance of drains.

## FAILURE OF ROADS

Failure of roads occurs if one of the major construction parts of the road structure does not function; it can then be defined as an occurrence which leaves the road impassable, or at least defective.

The main mechanisms of failure that can happen to a road are:

- total failure of a section of pavement
- erosion of the shoulder extending into the carriageway
- failure of the cross-carriageway drainage system.

The causes of each type of failure are of a different nature, and can be divided into four main groups:

- accidental obstructions
- inadequate construction
- drainage erosion
- drainage silting.

Table 6.1 provides a general view of the causes and their potential effects on earth and gravel roads. Most of these failures are caused by inadequate design or construction. It usually requires substantial work if the road is to be restored to a serviceable condition.

Both the pavement structure of a road and the volume of traffic which it carries are more relevant to the initial design of the road and to its periodic maintenance than they are to the calculation of its routine maintenance inputs.

**Table 6.1** Causes of failure and their effects on earth and gravel roads

| *Cause*<br>*Failures* | *Accidental obstruction* | *Inadequate construction* | *Drainage erosion* | *Drainage silting* |
|---|---|---|---|---|
| Pavement failure | | ○ inadequate earth work over weak subgrade<br>○ no stabilization on steep gradients<br>○ not sufficient camber | | |
| Shoulder/ carriageway erosion | ○ embankment failures blocking side drains | ○ gradients too steep with inadequate scour checks<br>○ no grass on shoulders or in base of side drains | ○ increasing erosion channels in shoulders or base of side drains | ○ overflow of silted side drains |
| Cross-carriageway drainage failure | ○ accidental blockage of culvert<br>○ accidental blockage of mitre drains | ○ inadequate culvert construction<br>○ inadequate mitre drains<br>○ inadequate gradient of culvert and culvert inlet/ outlet leading to silting | ○ erosion of side drains leading to collapse of culvert headwall | ○ blocked culvert<br>○ blocked mitre drains |

# Maintenance Systems

Maintenance contracts are allocated according to the basic maintenance operations:

- routine maintenance
- periodic maintenance
- emergency maintenance

It is necessary to distinguish between these maintenance operations to be able to:

- set priorities and plan maintenance work
- organize maintenance work
- quantify maintenance work
- estimate maintenance work for funding purposes
- allocate maintenance work to the maintenance personnel.

## ROUTINE MAINTENANCE

These are normally small-scale operations with limited resource requirements, usually performed at least once a year on a section of road. Routine maintenance consists of relatively unskilled activities, except for grading, which is a skilled operation.

It is necessary to define all routine maintenance activities clearly in order to be able to:

- quantify the activities
- instruct the activities to the maintenance personnel
- control and monitor the activities effectively.

The need for routine maintenance activities must be estimated, and the execution of the work must be planned and controlled.

**Table 6.2** Routine maintenance activities for unpaved roads

1. inspection and removal of obstacles
2. cleaning of drainage structures and their inlets and outlets (culverts, splashes, etc.)
3. repair of culvert headwalls, approaches and aprons of splashes
4. repair of culvert drains/off-shoot drains/catchwater drains and excavation to original sizes
5. cleaning of side drains and excavation to original size
6. cleaning of catchwater drains and excavation to original size
7. filling of pot-holes in carriageway
8. repair of shoulder and slope erosion
9. light reshaping of carriageway (camber formation, corrugation, ruts, etc.)
10. maintenance of erosion controls in drains
11. cutting of grass on shoulders and side drains
12. clearing bush

**Table 6.3** Routine maintenance activities for paved roads

1. inspection and removal of obstacles
2. cleaning of drainage structures and their inlets and outlets (culverts, splashes, etc.)
3. repair of culvert headwalls, approaches and aprons of splashes
4. repair of culvert drains/off-shoot drains/catchwater drains and excavation to original sizes
5. cleaning of side drains and excavation to original size
6. cleaning of catchwater drains and excavation to original size
7. patch and reshape shoulder (gravel shoulder)
8. patch surface edge
9. patch pot-holes, (including patching of local severe ruts/depressions)
10. seal cracks
11. sand off bleeding areas
12. bush clearing and grass cutting

Tables 6.2 and 6.3 provide checklists of routine maintenance activities for unpaved and paved roads respectively in approximate order of priority, although this can differ from case to case.

More road authorities are contracting out routine maintenance as it is difficult for them to carry out the work economically using direct labour (force account). Road authorities are usually centrally organized and it is difficult for them to control dispersed maintenance activities effectively. Small-scale contractors based in these locations should be able to offer a cheaper and more reliable service. Although routine maintenance contracts can never be a big business, they offer a continuous and steady workload, which is very attractive for a small-scale contractor.

## PERIODIC MAINTENANCE

These activities need to be carried out on a road, or a section of a road, after a number of years. They require extra resources from the client to implement.

Table 6.4 provides a checklist of periodic maintenance activities for which separate work units or sub-programmes could be established.

Periodic maintenance operations on gravel roads, especially regravelling, are usually contracted out. These contracts offer attractive possibilities to earn substantial amounts of money. To

**Table 6.4** Periodic maintenance activities

- heavy reshaping of road or road section (by labour, drag or towed grader)
- installation or reconstruction of small drainage structures
- rehabilitation of road or road section
- rehabilitation of major structures (bridges, drifts)
- reshaping and regravelling/resealing of road or road section
- provision of gravel stacks along the road to be used for routine maintenance activities

take on a regravelling contract the contractor needs access to expensive equipment which involves considerable costs; thus these contracts entail a large risk.

## EMERGENCY MAINTENANCE

These activities are required from time to time on a section of road whenever sudden and unforeseen damage occurs. In most cases this requires the deployment of additional resources.

By definition, emergency activities cannot be forecast during the client's annual maintenance needs assessment, so they cannot be priced directly into the annual maintenance contract. However, it is possible for the client to reserve a certain percentage of the routine maintenance funds for emergency cases, and include a provisional allocation for emergency maintenance in the contract. Normally the contractor will be required to include a schedule of unit rates in the contract submission, and these will provide the basis for calculating payments for any emergency work that may be authorised.

In order to plan realistically and organize the emergency activities when they happen it is necessary to identify the extent and kind of damage as fast and as exactly as possible.

**Table 6.5** Emergency maintenance activities

- Reconstruction or repair of damage to culverts / splashes resulting from washouts, erosion or breakage.
- Reconstruction or repair of damage to structures resulting from washouts, erosion, breakage or damage from high floods.
- Clearing of landslide, tree fall or rock fall.
- Reconstruction or repair of damage to a road section resulting from washout or serious erosion.
- Reconstruction or repair of damage to drainage systems resulting from serious silting up or erosion.
- Reconstruction or repair of damage to erosion protection resulting from serious washout, landslide, etc.

*Emergency assessment*

This assessment is often left to the routine maintenance contractor, as the contractor is required to inspect the road regularly and to inform the client immediately if any emergency arises.

The assessment should give sufficient information to:

- estimate the input of resources (manpower, material, equipment and tools)
- estimate the financial requirements
- develop an operational plan and organize the work immediately
- control and monitor the work.

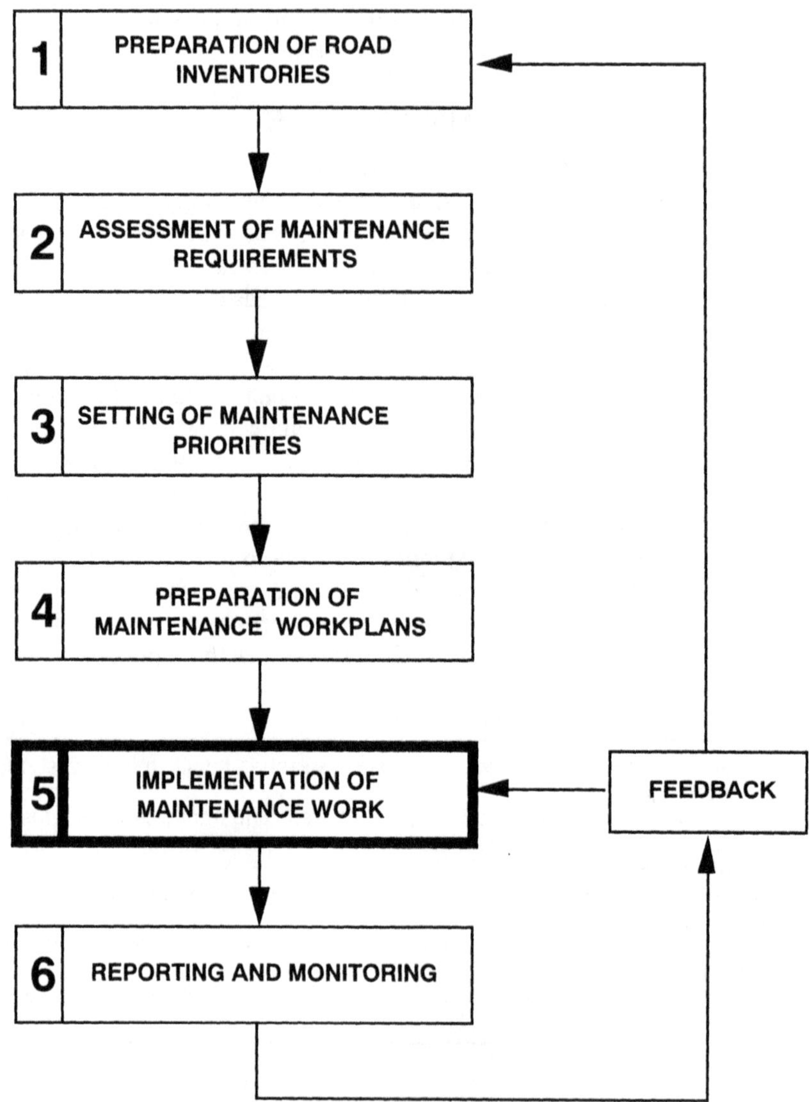


Figure 6.4 *Maintenance management cycle*

# Maintenance Management

## INTRODUCTION

All maintenance operations require careful planning, supervision and monitoring. The maintenance management cycle in Figure 6.4 shows the required activity phases in a logical sequence, from the client's viewpoint.

As a maintenance contractor you will not be dealing directly with the first three management phases. Your input will be expected to some degree for the preparation of workplans and your full involvement will start at the implementation phase (5).

In order to be able to understand fully your involvement at the implementation level, it is necessary to discuss the previous maintenance management phases briefly.

## ROAD INVENTORIES

Road Inventories are prepared by the client and should list and describe all important features of a road. The recorded data form the basic reference for all subsequent inspections and plans. A complete routine maintenance contract should also include an inventory of the road.

An example of a form used for such an inventory is shown in Table 6.6.

## ASSESSMENT OF MAINTENANCE REQUIREMENTS

The assessment of maintenance requirements has to be carried out by the client in order to identify the work that has to be done on a particular road. This assessment, which in principle is an inspection of the road's defects, is usually carried out on an annual basis. The assessments are used as basis for the preparation of detailed contracts and operational plans for routine and periodic maintenance.

This regular assessment is necessary:

- to make objective and quantified assessments of the condition of each road
- to review maintenance activities carried out since the previous inspection
- to determine routine maintenance activities for each road for the next period, and to issue contracts based on the actual maintenance needs
- to determine periodic maintenance activities to be carried out for the next period.

**Table 6.6** Road inventory (Example: Roads 2000 Kenya, MOPWH)

**Roads 2000** **ROAD INVENTORY** *(TICK OR COMPLETE BOXES AS APPROPRIATE)*

| DISTRICT | | MAINTENANCE ZONE | |
|---|---|---|---|

| ROAD CLASS + NO. | | START LOCATION | | END LOCATION | |
|---|---|---|---|---|---|

| SECTION NO. | | SECTION LENGTH | km |
|---|---|---|---|
| SECTION START IDENTIFIER | | SECTION END IDENTIFIER | |

**SURFACE TYPE:**

- Bitumen
- Gravel
- Earth

**PRINCIPAL SUBSOIL TYPE:**

- Red Coffee / Friable Clay
- Black Cotton / Expansive Clay
- Sandy Soil
- Volcanic Soil
- Stony / Natural Gravel
- Other

**GRADIENT**

% OF SECTION LENGTH

- ____ % Flat/Undulating: 0 - 3%
- ____ % Medium: 3 - 6%
- ____ % Hilly: > 6%

100 % TOTAL

**AVERAGE WIDTH**

____ M (TO NEAREST 0.5M)

**TRAFFIC FLOW (ADT)**

- 0 - 25
- 51 - 100
- 26 - 50
- > 100

- Manual Traffic Survey (MTS)
- Moving Observer Count (MOC)

Traffic Survey / Count Day: ____ ____ 199_

Day Month Year

**REGRAVELLING / RESEALING**

Date Last Regravelling / Resealing: ____ 19__

Month Year

Extent Regravelling / Resealing, Chainage: from km + ____ to km + ____

**GRAVEL SOURCES:**

| AVERAGE GRAVEL HAUL | km |
|---|---|

| NO. OF CULVERT LINES | |
|---|---|

**MAINTENANCE PRIORITY**

1 2 3 4 5

**REMARKS**

**STRUCTURES**

| CHAINAGE | TYPE OF STRUCTURE |
|---|---|
| | |
| | |
| | |
| | |
| | |
| | |
| | |
| | |

| DATE INVENTORY UPDATED | |
|---|---|

FORM PP 3 (F)

INTECH ASSOCIATES 9/91

## SETTING MAINTENANCE PRIORITIES

The annual assessment of maintenance requirements will identify the necessary work. Before the maintenance activities can be planned a most important intermediate step must be undertaken; the client has to allocate the limited resources available to achieve the desired result. Usually the financial resources available are not sufficient to carry out all the maintenance activities identified during the assessment. It is therefore necessary to set priorities for what maintenance activities and which road sections are most important.

Absolute priority is usually given to emergency maintenance activities, and a certain percentage of the available routine maintenance funds is set aside for it. This should then also be considered in the contract, and often the item 'miscellaneous' is meant to cover emergency maintenance activities. The client will instruct you, the contractor, how and when this item can be used. Remember always to ask for signed written instructions before you start any work that does not form part of the maintenance contract.

For the contractor, it is important to know that the client usually gives highest priority to routine maintenance of the drainage system. Neglected drainage work can quickly lead to a deterioration of the road. In simple terms, the first priority in routine maintenance is to make sure the rainwater runs off the road quickly and easily, causing as little damage as possible.

Other activities, such as grass cutting, bush clearing and slope maintenance are usually of lesser concern to the client and therefore receive a lower priority.

## PREPARATION OF WORK PLANS

For road maintenance, as for all other activities, it is always advantageous to prepare a work plan. There are two major types of road maintenance plan: long-term and short-term.

### *Long-Term Plans*

The long-term plans that are important for you are the general routine maintenance and the periodic maintenance plans. The general routine maintenance plans are based on an assessment of the maintenance requirements. The assessed requirements form the basis for the bill of quantities of your maintenance contract. Long-term maintenance plans are established by *the client*. The client knows what resources are available over a longer period of time. In this way the client also prepares long-term plans for periodic maintenance activities, such as regravelling. Based on data obtained from the annual assessment, the client is able to forecast when, for example, a further layer of gravel will be required.

*Example:*
*An assessment shows that the existing gravel layer of a certain road is presently on average 5cm thick, while the thickness of the layer when the road was gravelled 5 years ago, was 15cm. Based on that, the client assumes that the yearly gravel loss is about 2cm. So if nothing is done the road will have completely lost its gravel layer within 2–3 years. The client does not want the road to deteriorate completely so regravelling has to be undertaken one or, at the latest, two years from now.*

*Short-Term Plans or Operational Plans*
Short-term plans are the operational plans which *the contractor* has to prepare for the execution of the contract. The basis of these plans are the bill of quantities, the work standards and the time frame for the work as specified in the contract. Additional details influencing your plans can usually be obtained during the site inspection that you must always undertake before preparing the tender.

For routine maintenance, the operational planning process is discussed in detail in Chapter 7.

Operation plans are also necessary for regravelling. When asking you to forward a tender, the client usually also asks you to show him the work plan on which you have based the tender. It is therefore essential that you are capable of preparing such plans. We shall discuss operational plans for regravelling in Chapter 8.

## IMPLEMENTATION OF MAINTENANCE WORK

This is your actual contract work in the field! The next two Chapters, 7 and 8, look at the implementation of routine maintenance and regravelling contracts in detail.

## REPORTING AND MONITORING

Monitoring and reporting on the work carried out is important both for the client and the contractor:

- The client expects to get information on the quality and quantity of work carried out by the contractor.
- The contractor must receive information on resources actually spent and productivities achieved.

Therefore it is of vital importance to have good records of all site activities. *Without good records you cannot control your site.* The reports have to give you information on materials on site, labour and staff, as well as equipment and vehicles. Knowing how much money you are spending daily is one important aspect of reporting and monitoring. But you must also know how much money you are earning. That is, you need to measure all the work done for a certain period of time (per day, per week, etc.)

Merely measuring work done and recording resources used is not sufficient for a good monitoring process. These data must be compared with the targets set in your plans. A comparison of set targets against achieved work gives you a clear indication of your performance. Such an analysis must feed back into both the current work process and the planning of further work.

*Example:*
*If your site report from the gravelling site shows that the average daily output per tractor is only 6 instead of the targeted 8 loads of gravel, you have to check*

- *whether your assumption in the first place was correct, or*
- *what is the cause of the reduced output*

*and then decide an action to ensure you are not losing money on the contract. You may discover that there were not sufficient labourers appointed to load the gravel and hence the available labourers for this activity could not cope with the hauling capacity of the tractors. Your immediate feedback to the site staff would be to order a sufficient allocation of labour, enabling them to load as many tractor-loads as planned. For you as the company manager another feedback from this analysis could be that your site staff are not sufficiently trained to plan correctly for the daily implementation of your work plans.*

*Or were your work plans not detailed and accurate enough?*

## SAFETY MEASURES FOR WORK SITES

As a road work contractor you are directly responsible for the safety of your workers and the road users. Many road work activities are potentially dangerous and you have to minimize the risks by:

- ensuring that the necessary temporary traffic signs and protection are provided and correctly located on site for the duration of the works
- ensuring that all equipment and vehicles are parked off the carriageway or behind protective barriers and signs, when not in use

- ensuring that no materials are left in a dangerous location and that the road adjacent to the worksite is kept clean and swept of any debris arising from the work
- ensuring that all excavations are protected for the benefit of all road users, equipment and workers
- ensuring that all operators are trained in the operation of their equipment
- informing operators and labourers of the potential risks of, and procedures for, working with or close to machinery
- ensuring that traffic-control operations are properly carried out and that road users are not unnecessarily delayed
- ensuring that where work on the carriageway or shoulder remains unfinished overnight, proper warning lights are arranged and, if necessary, protected
- ensuring that all sites are left tidy and cleared of debris when the work is completed.

To protect your labourers who work on the road the following specific safety measures are recommended:

*Routine Maintenance Work on Low-Volume Roads*

- Either traffic signs 'Roadworks Ahead' or yellow/orange flags should be placed on the road shoulder before and after the site where the labourers are working.
- Labourers should wear yellow or orange safety vests or harnesses.

*Regravelling*

Wherever possible, diversions should be established so that traffic can be directed away from the road under regravelling. This will enable the job to be carried out more efficiently and safely. Properly installed road signs which clearly show the diversion are essential, and the road or road section under work needs to be blocked by adequate barriers and signals.

After the diversion has been established, and before work starts, warning signs, barriers and cones must be placed around the work area. Signs must be placed in the following order:

- 'Roadworks Ahead' signs should be placed 200m in front of the work area.
- 'Turn Left/Right' arrows should be placed 100m in front of the work area.
- Cones should be placed diagonally across the road to lead into the diversion.
- 'Keep Left/Right' arrows should be placed at the ends of the lines of cones.

- Barriers should be placed behind the lines of cones.
- ‘End of Restriction’ signs should be placed beyond the ends of the diversion.
- At night yellow lamps should also be used to mark the extent of the works at the diversions.

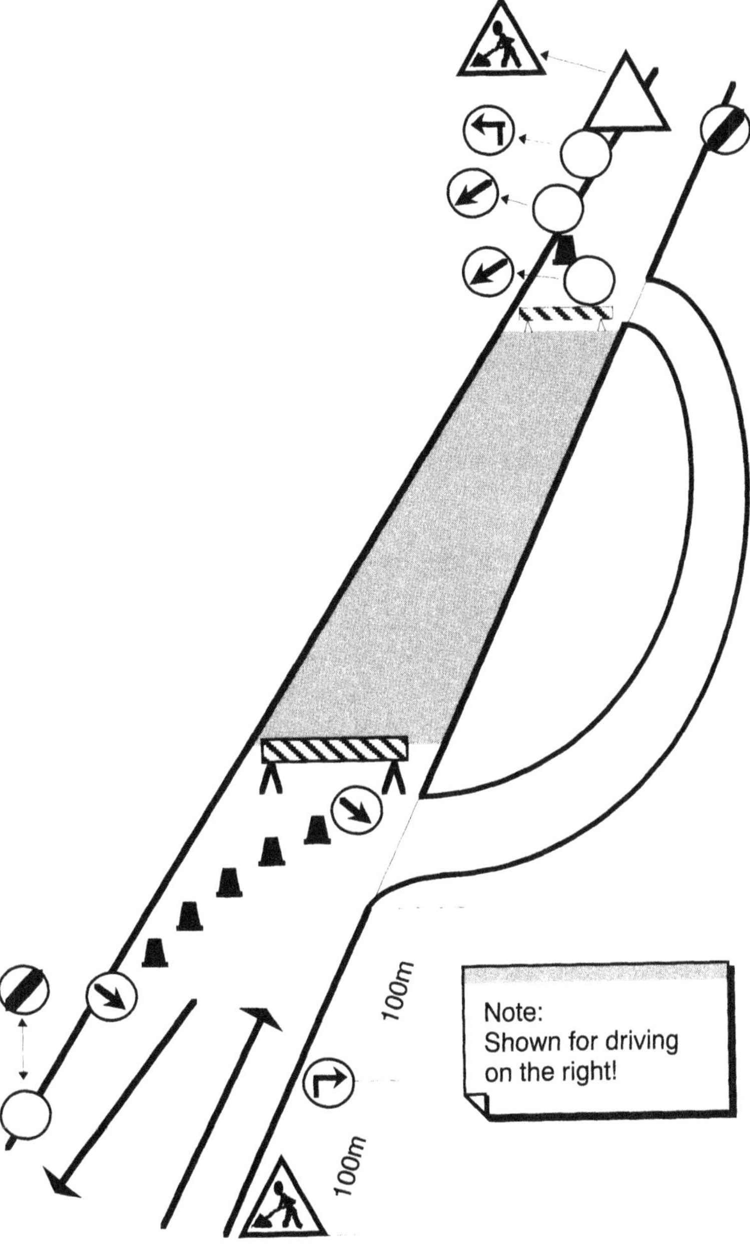


Figure 6.5 *Sign-posting for diversions*

However, very often it is not possible to divert traffic in developing countries as the road network is not dense enough and the construction of diversions is too expensive. Therefore roads are usually to be kept open to traffic. Before work starts, warning signs, barriers and cones must be placed around the work area. Work should be carried out on one side of the road at a time, allowing traffic to pass on the other.

Signs must be placed in the following order:

- 'Roadworks Ahead' signs should be placed 200m in front of the work area.
- 'Road Narrows' signs should be placed 100m in front of the work area.
- 'Speed Limit' signs should be placed at the start of the work area.
- Barriers should be placed at each end of the work area.
- Cones should be placed in a taper at the approaches to the work area and at a maximum spacing of 10m along the middle of the road next to the work area.
- 'End of Restriction' signs should be placed 50m beyond the work area.

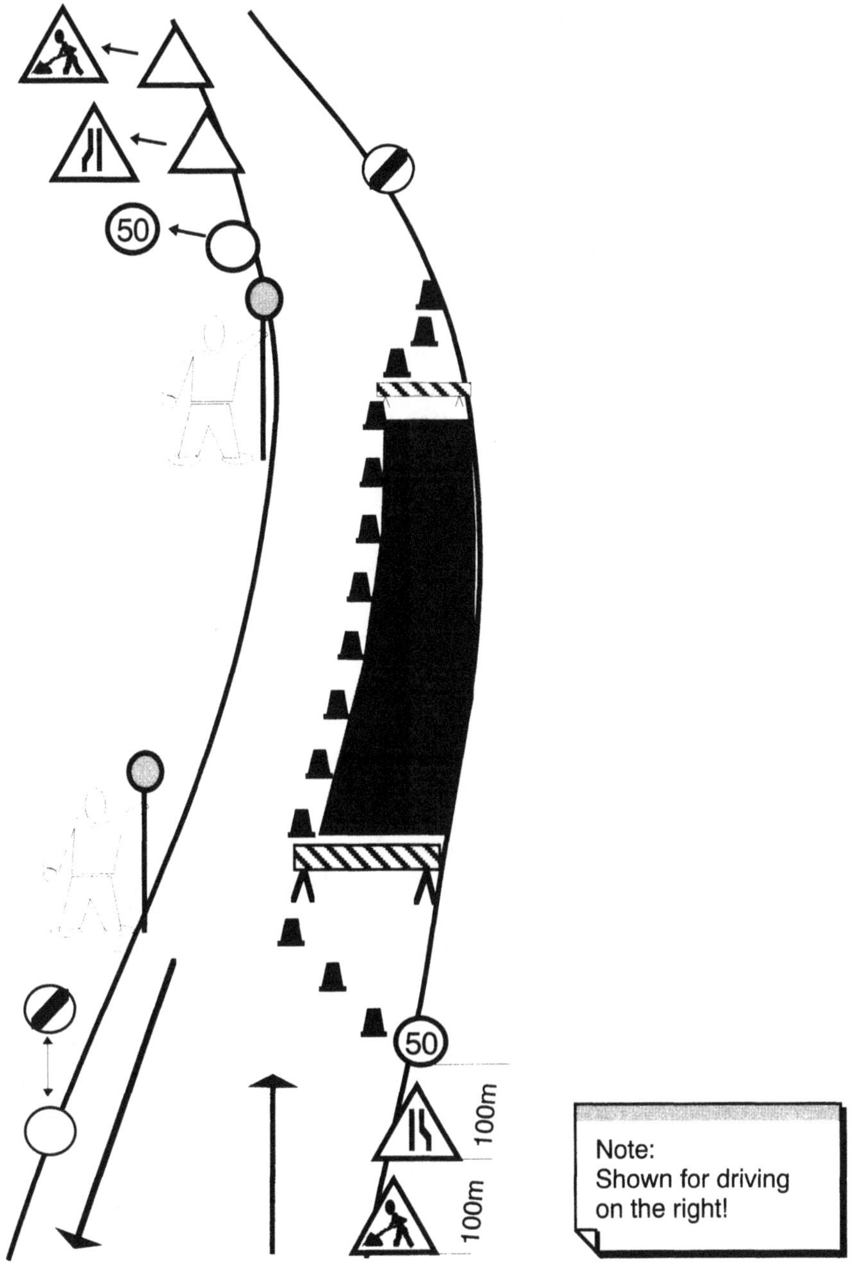


Figure 6.6 *Sign-posting for diversion of traffic to one side of the road*

# CHAPTER 7: ROUTINE MAINTENANCE

## Learning Objectives

After you have completed this chapter you should know:

- the basic operational methods of routine maintenance and how to organize work at site
- how to plan and monitor routine maintenance
- how to instruct and supervise your maintenance staff
- how to carry out all labour-based routine maintenance activities.

## Introduction

Following the general management aspects and the principles of road maintenance in Chapter 6 we can now discuss your involvement in implementing the maintenance work as a contractor. From the previous chapter you remember how your implementation on site fits into the client's overall management process (Figure 6.4).

While the long-term work plans for the entire network is the client's responsibility, the preparation of detailed operational plans (what to do on site) is the job of the contractor. How do you plan your job? Your two main sources of information are the contract documents and your experience.

The contract documents tell you how the client wants the job to be carried out. The contractor is fully responsible for achieving the standards specified within the time limits set, and at the quoted cost.

To be able really to benefit from your experience you need to follow up and document how all activities progress. Doing this regularly also means that you get a chance to correct mistakes at an early stage, and thereby save money.

This chapter consists of two sections. Part I, Planning, helps you to organize your routine maintenance work properly. It presents different organization models currently in use (length-person system, gang-system, etc.) and provides you with control sheets, facilitating the everyday planning and follow-up of your activities. Very often it is the way you organize your work that determines the level of profit you are able to achieve on your jobs.

The overall routine maintenance operation can be divided into single maintenance activities. Part Two of this chapter, Routine Maintenance Activities, describes the most important activities in detail. For each activity a suitable work method is suggested and a list of tools that are required is also provided.

# *Part I: Planning*

# ORGANIZING THE WORK

## GENERAL

A building site is usually compact and therefore relatively easy to control. However, routine maintenance activities have to be carried out along the entire road network and are therefore widely scattered.

The other problem with routine maintenance work is the nature of the work itself. On a construction site, where a new road is being constructed or rehabilitated, the work can be planned in detail within a defined time-frame. For routine maintenance this is not possible. Routine maintenance is not meant to produce a new product (the road) but to maintain an existing road at an acceptable standard. Routine maintenance is a continuous activity tackling the deterioration that can occur to a road at any time. Earth and gravel roads, especially, can deteriorate very quickly in tropical countries if not properly maintained.

The organization of work for routine maintenance has therefore to be rather different, as planning, supervision and control of work is a demanding task. The organizational set-up has to be based on the particular needs for routine maintenance, which simply means that work may have to be carried out anywhere on an entire road network at any time.

## SITE INSPECTION

Before you prepare a tender and corresponding work plans you must visit the road(s) and carry out a thorough site inspection. Make sure that you collect all the data necessary to prepare a realistic tender. To help you we propose the following checklist:

- What is the existing condition of the road in comparison to the required standards:
  - drainage (open or blocked?, sufficient mitre drains?, culverts in place?, drainage gradients correct?, scour checks in place?, etc.)
  - camber (is the gradient sufficient or does it need to be improved?)
  - roughness of carriageway
  - structures (are the structures in good condition or do they require repair work?)
  - is there any gravel left on the carriageway, and if so how much?
  - where are the trouble spots on the road and of what nature are they?
  - is there soil erosion along the road caused by the road and what must be done to avoid it?
- Does the road inventory and maintenance requirement assessment of the client correspond with your own assessment?
- What is the availability of skilled/unskilled labourers? Can you find labourers and gangleaders who have previously worked on the construction of the road and who are therefore experienced in labour-based road works?
- Are the potential labourers living close to the road and how long would be their average walking distance to their place of work?
- Can you get support from the local authorities to recruit and organize labour?
- Who are the people who 'matter' in the area of your operation?
- What is the transport distance from the company's base to the camp?
- What mode of transport is required and how many trips (tools, material, supervision, payments, etc.)?
- Is there water available along the road, a.) for drinking, b.) for compaction?

If the client does not provide you with a map of the road where all important features are marked it is recommended that you prepare your own sketch.

## TYPES OF CONTRACT

There are basically four different labour-based contract types for routine maintenance:

*1. Single lengthperson contract*
A contract for a defined section of a road (1 to 2km) is given to an individual.

*2. Petty contract (or labour group)*
A contract is given to a very small-scale contractor who in turn employs a small team (5 to 10 labourers) to maintain a defined section of a road (5 to 20km).

*3. Small-scale contract for a particular road*
A contract is given to a small to medium sized contractor who employs labourers to maintain a particular road or a longer road section (20 to 100km),

*4. Small-scale contract for a specified road network*
A contract is given to a small-scale contractor to maintain a specified road network, e.g. a full maintenance area covering all earth and gravel roads (100 to 300km of roads).

Contract types 3 and 4 are most likely to be of interest to a professional labour-based contractor. Before deciding to bid for a contract, you should consider:

- the supervisory capacity of your organization; how many supervisors do you have? What is their background?
- whether transport is available for all of them
- whether there are traditional limitations which affect the way in which work can be organized (for example, labourers may prefer working in a group)
- the distribution of people along the road (e.g. in villages or in scattered homes).

Based on these considerations you have to decide how to carry out your contract. There are a number of options to choose from.

## THE LENGTHPERSON SYSTEM

Figure 7.1 *The lengthperson system*

*System*
A labourer is appointed for each section of road, typically 1 to 2km in length. He or she is provided with hand tools and supervised regularly by a gangleader who monitors the condition of the road, directs operations, makes reports and authorizes payments for satisfactory work. A gangleader may be able to

supervise up to 10 labourers or 20km of roads. The labourer lives adjacent to the road and therefore does not require any transport. The labourer carries out all the routine maintenance activities on his or her particular road section. The task rate system is ideal for this sort of work.

The advantage of the lengthperson system is that a continuous maintenance of the entire road can be guaranteed at all times, and that one person can be made responsible for a specific road section. This system is particularly useful in high-rainfall areas where, for example, the opening of culverts and mitre drains needs to be carried out on the whole road length almost on a daily basis.

The disadvantage is that supervision has to take place on each and every section of road, which means that a gangleader or supervisor has to visit and talk to each labourer individually. The gangleader and supervisor have therefore to be mobile and the labourers have to be well trained so that they can work independently. For a large part of the time the gangleader is busy travelling from lengthperson to lengthperson.

*Transport*

Transport is required for the gangleader (bicycle) and for the contractor (pick-up) to supervise all maintenance work and to transport tools or other small materials (see also Chapter 4, Section 2).

*Tools*

Each lengthperson requires a standard set of hand tools:

- hoe
- shovel
- grass cutter
- bush knife (for wet areas where bush can encroach on to the road reserve)
- rake or spreader.

Two or three lengthpersons may share:

- wheelbarrow (to haul gravel from stockpiles or remove silt and organic material)
- earth rammer (for pot-hole filling).

The gangleader requires a basic set of measuring aids:

- tape measure
- ditch template
- spirit-level
- strings and pegs.

In some projects it may also be necessary to provide the gangleader with a long-handled shovel and trowel to clean out culverts.

## THE SINGLE GANG SYSTEM

Figure 7.2 *The single gang system*

*System*

A gang of 5 to 10 labourers, together with a gangleader, are appointed to maintain a single road of 5 to 20km. The gangleader monitors the condition of the road, directs operations, makes reports and authorizes payments for satisfactory work. The labourers live adjacent to the road but must be ready to walk longer distances than under the lengthperson system. The labourers carry out all routine maintenance and work under the direction of the gangleader. They start work at one end of the road and work their way up to the other end carrying out all necessary work. The task rate system can be applied for the entire gang (gang task) or tasks can be issued to individuals.

The advantage of this system is that supervision can be concentrated and planning of the work is easier. The gangleader can be utilized better as he or she is constantly with the team. In some cases it is possible that the gangleader not only supervises but also assists the other labourers in carrying out the work.

The disadvantage is that no continuous maintenance of the whole road section at all times is possible. For example, on a 15km-long road this could mean that one round of maintenance could last up to 6 months. Therefore this system is more effective in dry areas where frequent interventions to the drainage system are not necessary. Another disadvantage is that labourers often need to walk long distances until they reach their place of work.

*Transport*

Transport is required for the contractor (pick-up) to supervise all maintenance work and to transport tools or other small materials (see also Chapter 4, Section 2).

*Tools*

Each labourer requires a standard set of hand tools:

- hoe
- shovel
- grass cutter
- bush knife (for wet areas where bush can encroach on to the road reserve)
- rake or spreader.

As a gang they may share:

- 2 to 3 wheelbarrows (to haul gravel from stockpiles or remove silt and organic material)
- 1 to 2 earth rammers (for pot-hole filling)
- 1 bush saw.

The gangleader requires a basic set of measuring aids:

- tape measure
- ditch template
- spirit level
- strings and pegs.

In some projects it may also be necessary to provide the gang with a long-handled shovel and trowel to clean out culverts.

## THE GANG SYSTEM

Figure 7.3 *The gang system*

*System*

The maintenance activities are carried out by small labour gangs, each 4 to 5 labourers strong. Each gang is responsible for a section of road of approximately 6 to 10km. The gangleaders are also part of the labour force while monitoring the condition of the road, directing operations, making reports and authorizing payments for satisfactory work. The labourers live adjacent to the road but must be ready to walk some distance to work. The labourers carry out all routine maintenance work under the direction of gangleaders. They start work at one end of their section and work their way up to the other end, carrying out all necessary work. The task rate system can be applied for the entire gang (gang task) or tasks can be issued to individuals.

The advantage of this system is that supervision can be reduced and that planning of the work is easier. The gangleaders can be better utilized as they are constantly with the team and are also part of the labour force. This system allows frequent maintenance interventions on a particular road section and it is therefore possible to apply it in wet areas.

The disadvantage of this system is that some labourers still have to walk quite some distance until they reach their place of work. It is also not possible to provide continuous attention to the entire road at all times in high rainfall areas where very frequent work needs to be carried out.

*Transport*

Transport is required for the contractor (pick-up) to supervise all maintenance work and to transport tools or other small materials (see also Chapter 4, Section 2).

*Tools*

Each labourer requires a standard set of hand tools:

- hoe and shovel
- grass cutter
- bush knife (for wet areas where bush can encroach on to the road reserve)
- rake or spreader.

As a gang they may share:

- 1 to 2 wheelbarrows (to haul gravel from stockpiles or remove silt and organic material)
- 1 to 2 earth rammers (for pot-hole filling)
- 1 bush saw.

The gangleaders requires a basic set of measuring aids:

- tape measure
- ditch template and spirit-level
- strings and pegs.

In some projects it may also be necessary to provide the gangs with long-handled shovel and trowel to clean out culverts.

Unless the contract specifies a particular approach, you are free to choose whichever suits you best. Your aim is, of course, to achieve the required standards and quantities as specified in the contract, while making as much profit as possible.

*Example:*

*You are tendering for a routine maintenance contract for a gravel road of 34km length. The road is in relatively good shape and the drainage system is well established. The area where the road is located has only a little rainfall during the months of March and April. There are plenty of people living in the area where the road passes. In your firm you have only one or two potential gang-leaders at the time who could be deployed for road maintenance work. You also have limited transport facilities available, namely a pick-up for yourself and a lorry for the transport of materials. No specific way of organizing the maintenance is stipulated in the contract, so you are free to establish an organization that best suits your particular situation and the requirements of the work.*

*This means that you would set up two gangs of approximately eight labourers, each headed by a gangleader. The labourers would work three days a week, which leaves them with sufficient time to look after their farms. They would be in a position to keep the road up to the required standard as rains are to be expected only during a limited period of the year, and no continuous input along the entire road length is required. Drains can be opened before the rains start and the carriageway can be repaired during the dry season by making sure that at least two rounds of interventions are applied. This system would allow you to make the best use of your two available gangleaders and at the same time optimize the supervision by having them working together with the gangs. As labour is widely available it would be easy to select those living at approximately the middle of each section so that the walking distances would be acceptable.*

In most contracts you will be asked to carry out all routine maintenance activities, which includes the maintenance of the carriageway. However, you will remember that maintenance of the carriageway using labour-based methods may be possible only on roads which carry less than 50 vehicles per day. On roads with higher traffic, the carriageway may have to be graded, so you will need more equipment. The most economic form of maintenance grading is to use a tractor-towed grader or drag (see Figure 7.4). This equipment can be combined with any of the three systems discussed so far. However, you need to look carefully at the costs involved for such a mechanized operation. Small contracts will, in most cases, not permit full utilization of the equipment, so the cost involved will be too high to be economic (unless you can find useful work for the tractor on other sites).

*Example:*
*The 34km road described above may require four light grading cycles per year. This would keep your equipment busy for only approximately 50 to 60 working days per year. In this case you would do better to hire the equipment from an agent or another contractor. Alternatively, you could buy a grader, but hire a tractor from local farmers when you need it.*

## TOWED GRADING (DRAGGING)

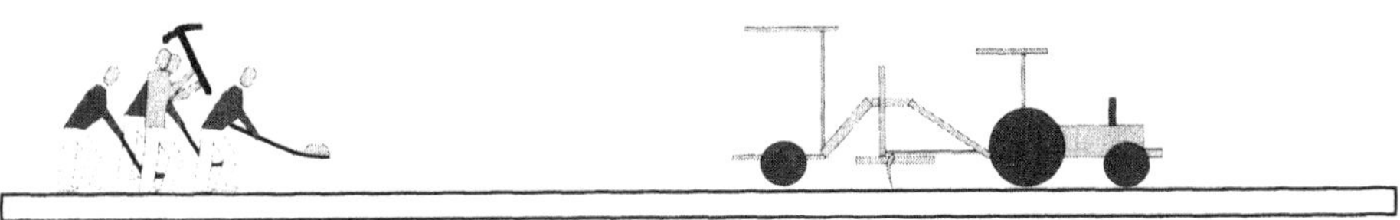

Figure 7.4 *Towed grading of carriageway*

*System*

For routine maintenance, towed grading is used to reshape the carriageway on high-trafficked roads which cannot be adequately maintained by labour only. The equipment operates in combination with any of the labour routine maintenance systems. Labour is used to carry out all off-pavement activities and for pot-hole filling where necessary. Dragging may be effective only on sandy soils and needs to be repeated very frequently. The task rate system can be applied for the entire gang (gang task) or tasks can be issued to individuals.

The advantage of this system is that the carriageway can easily be kept up to the required standard, where this is difficult for labour only. This applies mainly to high-trafficked roads (above 50 vehicles per day and on earth roads in high-rainfall areas with clay and silt).

The disadvantage of this system is that it depends on a piece of equipment which cannot be fully utilized for the particular contract. The equipment may be hired from other sources for the time required.

*Requirements*

- 1 Tractor 60hp with a light towed grader (2 tons), or
- 1 Tractor 4WD 100hp with heavy towed grader (5 tons)
- mechanical back-up service.

To operate this equipment two trained plant operators are required who will work under the supervision of the gangleader(s).

# Planning and Monitoring Routine Maintenance Work

The contractor's work can be specified in two different ways in a routine maintenance contract:

*Standard-based*
The contract defines the standards of specified road features that have to be maintained over a certain period, e.g. side ditch to be clean from silt, no erosion gullies, possible grass in ditch and on ditch-slopes to be kept short (50mm high) and ditch size to be in accordance with the standard size (60cm width and 35cm height). For this type of contract it is up to the contractor continuously to assess the road conditions and maintenance work that need to be done.

*Quantity-based*
The contract defines the exact quantity of work for each feature of the road to be maintained over a specified period of time, e.g. the contract is given for a period of three months and specifies ditch cleaning: clean silt from ditch to the specified ditch standard, quantity = 500m.

Planning and bidding for standard-based contracts is more difficult, since you have to estimate the quantities yourself in order to calculate how much work (and therefore cost) will be involved.

However, in both cases the client sets the priorities for the overall maintenance work, and you have to decide how to carry out the work in accordance with the contract. The operational planning of your work is necessary to make the best use of your available resources and to meet the requirements of the contract. The problem in planning road maintenance work is that you have to deal with ever-changing situations and conditions.

*Lump-sum-based*
The contract defines the standards of specified road features that have to be maintained and the total amount of money (lump-sum) the contract covers over a specified period of time. The lump-sum is based on cost assumptions made by the client. For example the client assumes the number of work-days which are required to carry out the work. Therefore the client also assumes the task rates for the various activities and allows for the other costs a contractor has, such as procurement and replacement of hand tools, transport for supervision and material, company costs and profit, etc. The client will then regularly instruct the contractor which activities are to be carried out over a certain period of

time. These instructions must be given in writing and cannot exceed the contract assumptions. It is therefore the right of the contractor to know these assumptions. The organization of the work is left to the contractor. Through effective and clever work organization the contractor can make additional profit.

*Example:*
*Your bid on the standard-based contract you were awarded was based on the condition assessment you made three months ago. You started work two months ago and now the rainy season has started. According to the contract the drainage should be open at all times. But now you realize that the ditches are silting up much faster than you had expected, which means that the work-load is more than you had estimated. There could be two reasons for this unexpected work-load; either your assessment was wrong, or you have made some mistakes when planning your operations. During the first two months you may have concentrated on activities like bush clearing, grass cutting and pot-hole filling instead of opening up all mitre drains and culverts. Now, when the rains start, the road cannot properly drain through these outlets and the ditches start to silt up. Either way, you will have* no basis for a claim *against the client on a standard-based contract (on a quantity-based contract, you would be paid more according to agreed unit rates, provided you were not at fault).*

As you can see, it is very important to plan maintenance activities carefully and to be aware of the different aspects that influence the operations. We therefore need to discuss maintenance priorities and planning principles in more detail.

## PRIORITIES FOR MAINTENANCE ACTIVITIES

As already mentioned in Chapter 6 the first priority in routine maintenance must be given to the drainage system, as neglected drainage work can quickly lead to deterioration of the road.

The main objective of routine maintenance on earth and gravel roads is to ensure that rainwater runs off the road as quickly as possible while causing as little damage as possible.

The road users would always argue that a smooth running surface has first priority and they are really not wrong, as the running surface is the part of the road which is actually used by vehicles. However, the road user may not realize that the condition of the surface depends very much on the function of the drainage system. It should also be quite obvious to a contractor that the work-load on the carriageway is reduced if you keep the drainage functioning well. Your aim as a contractor is, of course, always to carry out work as efficiently as possible, i.e. at the lowest possible cost, while meeting the required contract specifications. Therefore it does not make sense to clean out a ditch 5 times a year instead of twice simply because you have failed to open the mitre drains and culverts. (Even on a quantity-based contract, you would not be paid extra for work caused by your own bad management.)

When it comes to planning your activities you need to know what activity gives the best result at a particular time. Table 7.1 provides you with guidelines on when to carry out activities. However, these are only guidelines and you will have to develop your own priorities, depending on local conditions. You should therefore always investigate a road carefully before you start work. It is important that you get a clear picture of a road at all times, especially when you do not know what the particular road looks like during the rainy season. For example, you could talk to the local people who live along the road and ask them what the road looks like during the rainy season, where floods occur, where water is likely to form ponds, where vehicles get stuck, and so on. Such information helps you to detect problem areas and to set your own priorities.

**Table 7.1** Suggested priorities for routine maintenance activities

| *Season* | *Priority* | *Description* |
|---|---|---|
| Before rains | 1 | Clean culvert/splashes and bridges/drifts |
| | 2 | Clean mitre drains |
| | 3 | Clean side drains |
| | 4 | Repair side drain erosion and scour checks |
| During rains | 1 | Inspect and remove obstructions |
| | 2 | Clean culverts/splashes |
| | 3 | Clean side drains |
| | 4 | Clean mitre drains |
| | 5 | Repair side drain erosion and scour checks |
| | 6 | Repair erosion on shoulders |
| End of rains | 1 | Fill pot-holes and ruts in carriageway. |
| | 2 | Reshape carriageway |
| | 3 | Repair erosion on shoulders on back slopes and in drains |
| | 4 | Reinstate scour checks |
| | 5 | Cut grass |
| Dry season | 1 | Clear bush |
| | 2 | Repair structures |
| | 3 | Reshape carriageway |

## WORK PLANNING

The output specified in the contract provides your target, and your work plan is prepared so that you can meet that target.

Some contracts may even go so far as providing detailed operational plans, especially for a quantity-based contract. But in other cases you will need to develop your own work plans.

The data you need to produce a work plan are:

- the quantity of work to be carried out over a specified period of time
- the work standards to be achieved
- the resources you have available
- the productivity rates for each activity
- the chosen work system.

In a quantity-based contract, the quantity of work to be carried out is defined in the bill of quantities, and the period of time will also be specified in the contract. In a standard-based contract you must use your own experience (plus advice from others) to try to make the best possible estimate of the work involved.

The work standards to be achieved are set out in the contract document (often in an annexe). Where the standards are not clear to you, or are not specified in detail, you should ask the client for clarification. Make sure that the standards are clearly defined and written down before you commence work (so as to avoid disputes later).

The workplan will enable you to calculate the resources you need to carry out the work. Resources required for routine maintenance are:

- supervisors (gangleaders)
- labourers (mostly casual)
- hand tools
- measuring aids for gangleaders
- if necessary, simple equipment, such as a tractor and towed grader/drag (could be hired)
- supervision transport, such as a pick-up, a motorcycle or a bicycle
- transport for materials, such as a pick-up or a lorry (could be hired)

Estimated productivity rates are crucial both to planning and cost estimation. Do not rely on published productivity rates or figures given by other contractors – you should always check productivity on existing contracts and use these figures for planning and setting targets. Guidelines can assist you to develop your initial rates, but your experience is most important in developing your own productivity rates, and you always need to monitor carefully the actual achieved productivities at site. As a general guidance you can use the productivity table that is provided in Chapter 5, Table 5.5. Table 7.2 provides you with

**Table 7.2** Productivity standards for routine maintenance

| ACTIVITY | | UNIT | TASK DIFFICULTY 1 | 2 | 3 | 4 | NOTES |
|---|---|---|---|---|---|---|---|
| CLEAN CULVERTS + INLETS | | AS SHOWN | 4 CULVERTS/DAY | 1 CULVERT/ DAY | 2 DAYS/ CULVERT | 4 DAYS/ CULVERT | DIFFICULTY = SILT DEPTH IN CULVERT – 1. UP TO ¼. 2. ¼ TO ½. 3. ½ TO ¾. 4. OVER ¾.<br>TASKS FOR 600 DIA. CULVERTS WITH 7 RINGS |
| CLEAN CULVERT OUTFALLS | | M/DAY | 55 | 40 | 25 | | DIFFICULTY = SILT DEPTH – 1. UP TO 10CM. 2. 10 to 20CM. 3. OVER 20CM |
| REPAIR CULVERT HEADWALLS | | No/DAY | 7 | 4 | | | DIFFICULTY = TYPE OF REPAIR – 1. MINOR REPAIRS. 2. MAJOR REPAIRS |
| CLEAN MITRE DRAINS | | M/DAY | 60 | 45 | 30 | | DIFFICULTY = SILT DEPTH – 1. UP TO 10CM. 2. 10 TO 15CM. 3. OVER 15CM |
| CLEAN SIDE DRAINS | | M/DAY | WET AREAS 65<br>DRY SOFT SOIL 55<br>DRY HARD SOIL 30 | 45<br>40<br>23 | 30<br>30<br>18 | | DIFFICULTY = SILT DEPTH – 1. UP TO 10CM. 2. 10 TO 15CM. 3. OVER 15CM |
| REPAIR SCOUR CHECKS | | No/DAY | 5 | 7 | | | DIFFICULTY = TYPE OF SCOUR CHECK – 1. WOOD. 2. STONE |
| REPAIR SIDE DRAIN EROSION | | M/DAY | WET AREAS 100<br>DRY AREAS 100 | 80<br>50 | 60<br>23 | | DIFFICULTY = DEPTH OF EROSION – 1. UP TO 15CM. 2. 15 TO 30CM. 3. OVER 30CM |
| REPAIR SHOULDER EROSION | | M/DAY | 100 | 80 | 65 | | DIFFICULTY = DEPTH OF EROSION – 1. UP TO 10CM. 2. 10 TO 15CM. 3. OVER 15CM |
| GRASS PLANTING | | M/DAY | 100 | 80 | 65 | | DIFFICULTY = PLANTING WIDTH – 1. UP TO 0.5M. 2. 0.5 TO 1.0M. 3. OVER 1.0M |
| FILL POTHOLES IN CARRIAGEWAY | | W.BRWS/ DAY | 25 | 18 | 13 | 8 | DIFFICULTY = HAULING DISTANCE – 1. NO HAUL. 2. UP TO 100M. 3. 100M TO 200M. 4. OVER 200M |
| FILL RUTS IN CARRIAGEWAY | | M/DAY | WET AREAS 70<br>DRY AREAS 50 | 50<br>30 | 35<br>15 | 15<br>7 | DIFFICULTY = HAULING DISTANCE – 1. NO HAUL. 2. UP TO 100M. 3. 100 TO 200M. 4. OVER 200M |
| GRUB EDGE OF CARRIAGEWAY | | M/DAY | WET AREAS 270<br>DRY AREAS 190 | 200<br>120 | 130<br>70 | | DIFFICULTY = WIDTH OF GRUBBING – 1. UP TO 0.5M. 2. 0.5 TO 1.0M. 3. OVER 0.1M |
| RESHAPE CARRIAGEWAY* | | M/DAY | 70 | 50 | | | DIFFICULTY = TYPE OF RESHAPING – 1. LIGHT (UP TO 75MM). 2. HEAVY (OVER 75MM) |
| GRASS CUTTING | LIGHT | M/DAY | WET AREAS 425<br>DRY AREAS 310 | 260<br>230 | 190<br>170 | | DIFFICULTY = WIDTH OF GRASS CUTTING – 1. UP TO 1.0M. 2. 1.0 TO 2.0M. 3. OVER 2.0M |
| | DENSE | M/DAY | 310 | 240 | 175 | | |
| BUSH CLEARING | LIGHT | M/DAY | 425 | 260 | 190 | | DIFFICULTY = WIDTH OF BUSH – 1. UP TO 1.0M. 2. 1.0 TO 2.0M. 3. OVER 2.0M |
| | DENSE | M/DAY | 275 | 225 | 175 | | |

* ALL TASKS EXCEPT RESHAPING ARE MEASURED ALONG ONE SIDE OF THE ROAD ONLY

detailed productivity guidelines for common routine maintenance activities. However, use these guidelines with care and make sure you do not apply them blindly, as the local conditions you are working with may differ a lot.

It is advisable to use the task rate system for this type of work and the productivity rates have to be well developed for this purpose. (for details on the task rate system, see Chapters 5 and 10).

As already discussed earlier in this chapter, the work system you choose has a direct influence on the work execution and therefore also on the planning.

All these planning elements you will find scattered in different documents and tables. In order to have everything to hand we recommend that you prepare your own data-base for all routine maintenance activities. For each activity you could prepare a sheet which contains all the important information. This will help you in estimating, planning and supervising the work. Of course you need to update your sheets constantly based on your experience and the monitoring results you get during the execution of the work. An example of such an activity sheet is shown in Table 7.3.

*Overall Work Plan*

You should now be ready to prepare your operation plan. From the contract document you know the quantities for each activity. You also know from your estimates for the contract bid how many work-days you had calculated to finish each activity. The contract document should also tell you where to carry out the particular operation (locations). Based on this information you can prepare an overall operational plan which covers the entire contract duration. This plan looks very much like a labour schedule but also includes the quantities you plan to achieve. In order to make the plan a useful instrument for you to monitor the field performance, the form also allows you to record the achieved output so that you can compare it with the set targets. An example is shown in Table 7.4.

Use the form as follows:

1. Fill out the activities as listed in the contract, including the unit and quantity for each activity.
2. Check from your records (activity sheets) the relevant productivity rates you had assumed when preparing the tender and enter them into the column 'P. Rate'.
3. Calculate the work-days by dividing the quantity with the productivity rate and enter in the column 'WD'.
4. Calculate the total work-days for all contract activities and compare them with the work-days you actually have

**Table 7.3** Routine maintenance activity sheet

<table>
<tr><td>Activity<br>Clean mitre drains</td><td>Priority<br>1</td><td>Work definition<br>Clear all silt, debris, vegetation and ponded water and dispose of appropriately</td></tr>
<tr><td colspan="2">Standards<br>o to be clean from any silt, vegetation and rubbish<br>o minimum gradient 2% but preferably same as gradient of side ditch<br>o standard cross section:<br>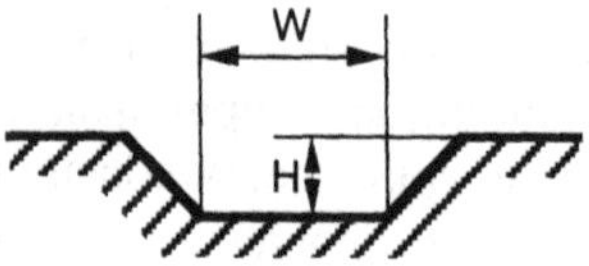
<br>W = Width = 60 cm<br>H = Height = 35 cm<br>side slopes = 1 : 1</td><td>Work method<br>o check correct gradient of drain<br>o re-establish or correct gradient if necessary<br>o where the drain is not the standard size anymore, set out correct size using strings and pegs<br>o clean mitre drain from silt, vegetation and/or rubbish and ensure standard shape<br>o re-establish block-off where necessary</td></tr>
<tr><td>Manpower<br>o Gangleader for setting out and supervision<br>o Labourer to clean</td><td>Tools and equipment<br>o standard hand tool set for labourers<br>o 1 wheelbarrow<br>o Gangleaders measuring aids</td><td>Material<br>o pegs and strings</td></tr>
<tr><td colspan="3">Productivity<br>clean mitre drains: fully silted = 20 metres / day / labourer<br>2/3 silted = 30 metres / day / labourer<br>1/3 silted = 50 metres / day / labourer</td></tr>
</table>

available during the contract period. These you will get by multiplying the number of labourers you had planned for the job by the available working days during this period. (Check the calendar for public holidays). Should there be a difference in the total work-days required and those available, then you must try to balance the labour input to match the required work-days.

5. Now you have to allocate the work-days to the weeks. It is important that the total work-days allocated to each week is not more than the total number of working days per week times the number of labourers you have in the field. (In our example this would normally be 16 labourers times 5 days =

**Table 7.4** Maintenance workplan and report



## ROUTINE MAINTENANCE PLAN AND REPORT FOR 3 MONTHS

**Road:** **Length:** Km **Period: from** **to**

**No. of Labourers:** **Avail. Working Days:** **Avail. Work-days:**

| Activity | Unit | Quantity | P. Rate | WD | | Week 1 | | Week 2 | | Week 3 | | Week 4 | | Week 5 | | Week 6 | | Week 7 | | Week 8 | | Week 9 | | Week 10 | | Week 11 | | Week 12 | | Total | |
|---|---|---|---|---|---|---|---|---|---|---|---|---|---|---|---|---|---|---|---|---|---|---|---|---|---|---|---|---|---|---|---|
| | | | | | | WD | Q | WD | Q | WD | Q | WD | Q | WD | Q | WD | Q | WD | Q | WD | Q | WD | Q | WD | Q | WD | Q | WD | Q | WD | Q |
| | | | | | Pl | | | | | | | | | | | | | | | | | | | | | | | | | | |
| | | | | | Ach | | | | | | | | | | | | | | | | | | | | | | | | | | |
| | | | | | Pl | | | | | | | | | | | | | | | | | | | | | | | | | | |
| | | | | | Ach | | | | | | | | | | | | | | | | | | | | | | | | | | |
| | | | | | Pl | | | | | | | | | | | | | | | | | | | | | | | | | | |
| | | | | | Ach | | | | | | | | | | | | | | | | | | | | | | | | | | |
| | | | | | Pl | | | | | | | | | | | | | | | | | | | | | | | | | | |
| | | | | | Ach | | | | | | | | | | | | | | | | | | | | | | | | | | |
| | | | | | Pl | | | | | | | | | | | | | | | | | | | | | | | | | | |
| | | | | | Ach | | | | | | | | | | | | | | | | | | | | | | | | | | |
| | | | | | Pl | | | | | | | | | | | | | | | | | | | | | | | | | | |
| | | | | | Ach | | | | | | | | | | | | | | | | | | | | | | | | | | |
| | | | | | Pl | | | | | | | | | | | | | | | | | | | | | | | | | | |
| | | | | | Ach | | | | | | | | | | | | | | | | | | | | | | | | | | |
| | | | | | Pl | | | | | | | | | | | | | | | | | | | | | | | | | | |
| | | | | | Ach | | | | | | | | | | | | | | | | | | | | | | | | | | |
| | | | | | Pl | | | | | | | | | | | | | | | | | | | | | | | | | | |
| | | | | | Ach | | | | | | | | | | | | | | | | | | | | | | | | | | |
| | | | | | Pl | | | | | | | | | | | | | | | | | | | | | | | | | | |
| | | | | | Ach | | | | | | | | | | | | | | | | | | | | | | | | | | |
| | | | | | Pl | | | | | | | | | | | | | | | | | | | | | | | | | | |
| | | | | | Ach | | | | | | | | | | | | | | | | | | | | | | | | | | |
| Total | | | | | Pl | | | | | | | | | | | | | | | | | | | | | | | | | | |
| | | | | | Ach | | | | | | | | | | | | | | | | | | | | | | | | | | |

P. Rate = Productivity Rate (Task Rate) WD = Work-days Q = Quantity Pl = Planned Output Ach = Achieved Output

80 work-days). Fill your allocation into the boxes 'PL – WD' which means 'Planned Work-days'. Now calculate the quantity of work to be achieved during this particular week by multiplying the Productivity Rate by the Planned Work-days. Fill the result into box 'Q' which means 'Quantity'. Also make sure that you concentrate on as few activities as possibly over a period of one or two weeks. This makes it easier to supervise and control.

6. Total all the WD columns and check that their sum corresponds with the available work-days. The totals in the Total column on the right-hand side of the form have to agree with the Quantity and WD figures on the left-hand side. The totals in the Total row at the bottom of the form have to agree with the available work-days per week and, in the last box, with the total work-days available during the contract.

The form is designed to cover a period of three months. If you should have a contract for more than three months then you can simply add another form.

At the end of every week when you get the field reports back you can fill out the achieved work-days and quantities into the respective boxes and compare the results with the planned targets.

Table 7.5 is an example of a completed form based on the contract example on page 137. It is assumed that the client has awarded to you a contract which considers the actual maintenance needs for a three month period. According to your calculations you would require 540 work-days to complete the job. As you are offering employment to labourers living next to the road who are also farmers, you cannot give them full-time employment. At the same time the client wants you to make sure that the road receives continuous maintenance attention over the full contract period. You therefore decide to have two gangs each working every second day only (three days a week).

Note: For routine maintenance activities it is not possible to establish exact work plans as the situation can and will change during the contract period. Therefore it is important that you use your overall plan as a planning guideline which helps you to prepare the weekly work plans for the gangleaders. It is also important that your weekly inspections and observations are taken into account when preparing the next week's plans. The overall planning form also helps you to keep track of the overall consumption of work-days and achieved output in comparison to the contract BQ.

*Weekly Field Instructions and Report*

The overall plan is also used by you to prepare daily site instructions for the different gangleaders or even for individual

**Table 7.5** Maintenance workplan and report (filled example)

**ROUTINE MAINTENANCE PLAN AND REPORT FOR 3 MONTHS** **Road:** *Bonoki-Maruba* **Length:** *34.000* Km **Period: from** *1.1.* **to** *31.3.96*

**No. of Labourers:** *18* **Avail. Working Days:** *30* **Avail. Work-days:** *540*

| Activity | Unit | Quantity | P. Rate | WD | | Week 1 | | Week 2 | | Week 3 | | Week 4 | | Week 5 | | Week 6 | | Week 7 | | Week 8 | | Week 9 | | Week 10 | | Week 11 | | Week 12 | | Total | |
|---|---|---|---|---|---|---|---|---|---|---|---|---|---|---|---|---|---|---|---|---|---|---|---|---|---|---|---|---|---|---|---|
| | | | | | | WD | Q | WD | Q | WD | Q | WD | Q | WD | Q | WD | Q | WD | Q | WD | Q | WD | Q | WD | Q | WD | Q | WD | Q | WD | Q |
| *Inspection and Removal of Obstructions* | *m* | *34 000* | *--* | *24* | Pl | *2* | | *2* | | *2* | | *2* | | *2* | | *2* | | *2* | | *2* | | *2* | | *2* | | *2* | | *2* | | *24* | *34 000* |
| | | | | | Ach | | | | | | | | | | | | | | | | | | | | | | | | | | |
| *Clean Culverts* | *No* | *102* | *2* | *51* | Pl | *6* | | *6* | | *6* | | *6* | | *6* | | *6* | | *6* | | *6* | | *3* | | | | | | | | *51* | *102* |
| | | | | | Ach | | | | | | | | | | | | | | | | | | | | | | | | | | |
| *Clean Side Drains* | *m* | *14 500* | *50* | *290* | Pl | *30* | | *30* | | *24* | | *30* | | *24* | | *22* | | *26* | | *22* | | *23* | | *21* | | *16* | | *22* | | *290* | *14 500* |
| | | | | | Ach | | | | | | | | | | | | | | | | | | | | | | | | | | |
| *Repair Scour Checks* | *No* | *56* | *5* | *11* | Pl | *1* | | *1* | | *1* | | *1* | | *1* | | *1* | | *1* | | *1* | | *1* | | *1* | | *1* | | | | *11* | *56* |
| | | | | | Ach | | | | | | | | | | | | | | | | | | | | | | | | | | |
| *Clean Mitre Drains* | *m* | *1850* | *60* | *31* | Pl | *6* | | | | | | | | *7* | | | | | | *9* | | *4* | | *5* | | *0* | | *0* | | *31* | *1850* |
| | | | | | Ach | | | | | | | | | | | | | | | | | | | | | | | | | | |
| *Fill Pot-holes* | *No* | *120* | *20* | *6* | Pl | | | | | *2* | | | | | | *2* | | | | | | *2* | | | | | | | | *6* | *120* |
| | | | | | Ach | | | | | | | | | | | | | | | | | | | | | | | | | | |
| *Grub Carriageway Edge* | *m* | *9200* | *250* | *37* | Pl | | | *4* | | | | | | | | *10* | | | | | | | | *10* | | *10* | | *3* | | *37* | *9200* |
| | | | | | Ach | | | | | | | | | | | | | | | | | | | | | | | | | | |
| *Repair Shoulder - Slope Erosion* | *m* | *250* | *50* | *5* | Pl | | | | | | | | | | | *2* | | | | *2* | | | | *1* | | | | | | *5* | *250* |
| | | | | | Ach | | | | | | | | | | | | | | | | | | | | | | | | | | |
| *Cut Grass* | *m2* | *22 050* | *350* | *63* | Pl | | | | | *10* | | | | *5* | | | | *10* | | | | *10* | | | | *10* | | *18* | | *63* | *22 050* |
| | | | | | Ach | | | | | | | | | | | | | | | | | | | | | | | | | | |
| *Clear Bush* | *m2* | *4 400* | *200* | *22* | Pl | | | *2* | | | | *6* | | | | | | | | *3* | | | | *5* | | *6* | | | | *22* | *4400* |
| | | | | | Ach | | | | | | | | | | | | | | | | | | | | | | | | | | |
| | | | | | Pl | | | | | | | | | | | | | | | | | | | | | | | | | | |
| | | | | | Ach | | | | | | | | | | | | | | | | | | | | | | | | | | |
| | | | Total | *540* | Pl | *45* | | *45* | | *45* | | *45* | | *45* | | *45* | | *45* | | *45* | | *45* | | *45* | | *45* | | *45* | | *540* | |
| | | | | | Ach | | | | | | | | | | | | | | | | | | | | | | | | | | |

P. Rate = Productivity Rate (Task Rate) WD = Work-days Q = Quantity Pl = Planned Output Ach = Achieved Output

lengthpersons. Site instructions will cover a period of one or two weeks at most. In Table 7.6 there is an example of a gangleader's blank instruction and report form. The form also specifies the locations where the activities have to be carried out. The aim is to concentrate on as few activities as possible in a given period of time, because this makes supervision and control easier. It is important that these instructions are carefully explained to the gangleaders.

**Table 7.6** Gangleader's weekly instructions and report (blank)

| COMPANY: | GANGLEADER'S INSTRUCTIONS | ROUTINE MAINTENANCE |
|---|---|---|

| ROAD NAME: | ROAD NO: | GANG REF. | GANGLEADER'S NAME: |
|---|---|---|---|

| INSTRUCTION FOR WEEK: | BY : | DATE OF INSTRUCTION |
|---|---|---|

| GANG SIZE: NO A | WORKING DAYS THIS PERIOD: NO B | TOTAL WORK-DAYS AVAILABLE: NO AxB = C |
|---|---|---|

| ACTIVITY | LOCATION FROM – TO G | PLANNED WORK-DAYS D | TASK RATE E | PLANNED OUTPUT DxE = F | ACHIEVED WORK-DAYS H | ACHIEVED OUTPUT K |
|---|---|---|---|---|---|---|
| | | | | | | |
| | | | | | | |
| | | | | | | |
| | | | | | | |
| | | | | | | |
| | | | | | | |
| | | | | | | |
| TOTAL WORK-DAYS | planned | | | achieved | | |

Fill the form as follows:

1. Enter all the details on the header of the form. Prepare a separate form for each gang.
2. Extract the activities you had planned for the particular week from the Overall Plan and enter them into the 'Activity' column.
3. Enter in box 'A = Gang Size' the number of labourers you have in the particular gang.

4. Enter in box 'B = Working Days This Period' the number of working days that are available during the week (calendar days).
5. Calculate in box 'C = Total Work-days' the work-days you have available during this particular week, by multiplying the Gang Size (A) by The Working Days (B).
6. Extract from the overall plan the planned work-days for the particular activities and enter it into column 'D = Planned Work-days'.
7. Extract from the overall plan the Productivity Rate (Task Rate) for the particular activity and enter it into the column 'E = Task Rate'.
8. Calculate in column 'F = Planned Output' the total work to be achieved under this activity by multiplying the Planned Work-days (D) by the Task Rate (E).
9. Now you know the planned output and you can allocate the work to the location where it needs to be carried out. Check from the contract document the exact location start for the particular activity and enter this into column 'G = Location'. This should usually be a chainage, but if it is not, you may enter the description of the location, e.g. junction of road X. Now you add to this location the length of the section you can cover with the planned output you had calculated in column F.
10. The form is now ready to be given to the gangleader. Make sure you explain all the details very carefully to the gangleader. It will be best if you do that on site and show him or her the different activities to be carried out, and their locations. At the end of the week the gangleader will enter the work-days actually spent by his or her gang on each activity into column 'H = Achieved Work-days'. He or she therefore needs to keep a daily record on a separate sheet of paper from which he or she can transfer the data at the end of the week.
11. Column 'K = Achieved Output' can either be filled out by you when you inspect the achieved work at the end of the week, or by the gangleader if you trust him or her fully. The achieved output should not be a multiplication of column H by column E, but should actually reflect the measured output.
12. The results of the columns 'H = Achieved Work-days' and 'K = Achieved Output' can now be transferred to your overall plan.

The gangleader then reports the achieved work using the same form, which makes it easier for you to monitor the site performance. A completed example of such a form is shown in Table 7.7.

**Table 7.7** Gangleader's weekly instructions and report (filled example)

| COMPANY: *TECHNOPLUS* | GANGLEADER'S INSTRUCTIONS | ROUTINE MAINTENANCE |
|---|---|---|

| ROAD NAME: | ROAD NO: | GANG REF. | GANGLEADER'S NAME: |
|---|---|---|---|
| *BONOKI - MARUBA* | *E22* | *E22/1* | *Jacob Ondieki* |

| INSTRUCTION FOR WEEK: | BY : | DATE OF INSTRUCTION |
|---|---|---|
| *No6(4.7.-8.7.95)* | *Peter Foreman* | *3.7.95* |

| GANG SIZE: A | WORKING DAYS THIS PERIOD: B | TOTAL WORK-DAYS AVAILABLE AxB = C |
|---|---|---|
| *8* NO | *3* NO | *24* NO |

| ACTIVITY | LOCATION FROM – TO G | PLANNED WORK-DAYS D | TASK RATE E | PLANNED OUTPUT DxE = F | ACHIEVED WORK-DAYS H | ACHIEVED OUTPUT K |
|---|---|---|---|---|---|---|
| *Inspection and Removal of Obstructions* | *0.000 - 17.000* | *1* | - | - | *1* | - |
| *Clean Culverts* | *0.000 - 17.000* | *4* | *2No* | *8NO* | *4* | *7No* |
| *Clean side Drains* | *0.000 - 17.000* | *13* | *50m* | *650m* | *12* | *600m* |
| *Repair Scour Checks* | *0.000 - 17.000* | *1* | *5No* | *5No* | *1* | *5No* |
| *Cut Grass* | *0.000 - 17.000* | *5* | $350m^2$ | $1750m^2$ | *5* | $1750m^2$ |
| | | | | | | |
| | | | | | | |
| TOTAL WORK-DAYS | planned | *24* | | achieved | *23* | |

The achieved output reported in the gangleader's forms are then transferred to your overall work plan which now allows you to compare the planned targets with the actual work achieved.

These two forms are suggestions and are meant only to give you guidance in developing your own forms which suit you and your contract best. They are seen as the minimum requirements on planning and reporting forms for routine maintenance.

In addition to the two forms it is essential to maintain a muster-roll on site. This is a daily record of each labourer's attendance . The muster-roll is filled out by the gangleader every day. He or she records whether the labourers reported for duty or not, and whether they completed their task or not. The muster-roll is collected by you together with the site report at the end of the month. The work-days reported in the site report must correspond with the days reported in the muster-roll. It is important that the gangleaders keep this record very accurately and you therefore need to teach them carefully how to use it.

Based on the muster-roll you prepare each labourer's salary.

Procedure to fill out the muster-roll:

1. A new muster-roll is opened each month (a muster-roll can also be kept for shorter periods if necessary).
2. Fill the headings on the muster-roll and list the names of all labourers of the gang/site.
3. Each morning make a vertical line in the appropriate box for all people who have reported for work. If a person does not report, complete the line to make an A (Absent).
4. When a task has been completed and approved by the gang-leader, the worker can be released and a P (Present) is made out of the line.
5. If a worker does not complete his or her task and leaves the site without permission, enter A (Absent) against the his or her name.
6. If the worker has not completed his or her task because of unforeseen circumstances but is considered to have done a full day's work, enter P (Present) against his or her name.
7. If a worker cannot complete a task on the particular day because he or she feels unable to complete it (sickness, etc.) the worker should be given the chance to complete the task on the following day, but will be paid for one day only.
8. Mark non-work days with an X.
9. At the end of the month or reporting period the muster-roll should be checked against the daily site reports.

Tables 7.8 and 7.9 show a blank and a completed muster-roll, respectively.

When planning routine maintenance work, an additional aspect to consider is emergency maintenance. Some of the contracts may include an item called 'Miscellaneous' which is kept in the contract for any work that may be required but which cannot be planned. The amount included may be small, and is usually expressed as a lump-sum. The contract specifies the activities for which this item can be used. Small emergency activities that are identified during the daily inspections by the gangleader can then be carried out using this item, as it is important that the maintenance team reacts immediately to such needs. Small emergency activities could include:

- small landslide blocking the road
- drift/splash bed washed out, eroded by floods
- backfill at a structure washed out
- culvert washed out or broken, etc.

**Table 7.8** Muster-roll

| COMPANY: | MUSTER-ROLL | ROUTINE MAINTENANCE |
|---|---|---|

| ROAD NAME / NO: | MONTH: | GANG REF. | GANGLEADER'S NAME: |
|---|---|---|---|

| PAY NO | NAME | CAT | M | T | W | T | F | S | S | M | T | W | T | F | S | S | M | T | W | T | F | S | S | M | T | W | T | F | S | S | M | T | W | T | F | S | S | TOTAL DAYS |
|---|---|---|---|---|---|---|---|---|---|---|---|---|---|---|---|---|---|---|---|---|---|---|---|---|---|---|---|---|---|---|---|---|---|---|---|---|---|---|
| | | | | | | | | | | | | | | | | | | | | | | | | | | | | | | | | | | | | | | |
| | | | | | | | | | | | | | | | | | | | | | | | | | | | | | | | | | | | | | | |
| | | | | | | | | | | | | | | | | | | | | | | | | | | | | | | | | | | | | | | |
| | | | | | | | | | | | | | | | | | | | | | | | | | | | | | | | | | | | | | | |
| | | | | | | | | | | | | | | | | | | | | | | | | | | | | | | | | | | | | | | |
| | | | | | | | | | | | | | | | | | | | | | | | | | | | | | | | | | | | | | | |
| | | | | | | | | | | | | | | | | | | | | | | | | | | | | | | | | | | | | | | |
| | | | | | | | | | | | | | | | | | | | | | | | | | | | | | | | | | | | | | | |
| | | | | | | | | | | | | | | | | | | | | | | | | | | | | | | | | | | | | | | |
| | | | | | | | | | | | | | | | | | | | | | | | | | | | | | | | | | | | | | | |

| CERTIFIED CORRECT (GANGLEADER) | CHECKED BY: | DATE |
|---|---|---|
| DATE __ __ 199_ (Day, Month, Year) | CHECKED BY: | DATE |

It is therefore always good to allow a few work-days during the contract period for such occurrences.

Where bigger inputs are required but which are not covered by your contract, you need to inform the client immediately so that he or she can prepare for the necessary arrangements. It is therefore important that you train your gangleaders to inspect the road carefully on a daily basis and to report to you any major occurrence. We also recommend that in every case of emergency work you prepare a detailed assessment. This can be done using a sheet of paper on which you record the exact location of the occurrence, describe the nature of the damage and estimate the work required to solve the problem. Also add the date and your signature to it and retain a copy for yourself, while giving the original to the client.

**Table 7.9** Muster-roll (filled example)

| COMPANY: Technoplus | MUSTER-ROLL | ROUTINE MAINTENANCE |
|---|---|---|

| ROAD NAME / NO: | MONTH: | GANG REF. | GANGLEADER'S NAME: |
|---|---|---|---|
| Banaki-Maruba (E 22) | April 1995 | E22/1 | Jacob Ondieki |

| PAY NO | NAME | CAT | M | T | W | T | F | S | S | M | T | W | T | F | S | S | M | T | W | T | F | S | S | M | T | W | T | F | S | S | M | T | W | T | F | S | S | TOTAL DAYS |
|---|---|---|---|---|---|---|---|---|---|---|---|---|---|---|---|---|---|---|---|---|---|---|---|---|---|---|---|---|---|---|---|---|---|---|---|---|---|---|
| 22/1 | Joseph Kamau | | P | P | P | P | P | X | X | P | P | P | P | P | X | X | P | P | P | P | P | X | X | P | P | P | P | P | X | X | | | | | | | | 20 |
| 22/2 | Peter Nduriri | | P | A | P | P | P | X | X | P | P | P | P | P | X | X | P | P | P | P | P | X | X | P | P | A | P | P | X | X | | | | | | | | 18 |
| 22/ | Wanjiru Kirika | | P | P | P | P | P | X | X | P | P | P | P | P | X | X | P | P | P | P | P | X | X | P | P | P | P | P | X | X | | | | | | | | 20 |
| 22/3 | Elisabeth Omango | | P | P | P | P | P | X | X | A | A | P | P | P | X | X | P | P | P | P | A | X | X | P | P | P | P | P | X | X | | | | | | | | 17 |
| 22/5 | Nathan Ndirita | | P | P | P | P | P | X | X | P | P | P | P | P | X | X | P | P | P | P | P | X | X | P | P | P | A | A | X | X | | | | | | | | 18 |
| 22/ | Suzanne Wanjoki | | P | P | P | A | P | X | X | P | P | P | P | P | X | X | P | P | P | P | P | X | X | P | P | P | P | P | X | X | | | | | | | | 19 |
| 6 | | | | | | | | | | | | | | | | | | | | | | | | | | | | | | | | | | | | | | |
| | | | | | | | | | | | | | | | | | | | | | | | | | | | | | | | | | | | | | | |
| | | | | | | | | | | | | | | | | | | | | | | | | | | | | | | | | | | | | | | |
| | | | | | | | | | | | | | | | | | | | | | | | | | | | | | | | | | | | | | | |

| CERTIFIED CORRECT (GANGLEADER) | CHECKED BY: | DATE |
|---|---|---|
| Jacob Ondieki | CHECKED BY: | DATE |
| DATE 30 04 1996 (Day Month Year) | | |

## WORK SUPERVISION AND MONITORING

As for any other contract work, supervision and monitoring by the contractor himself is very important and has the following purposes:

- to assure the work is carried out in accordance with the contract specifications and standards
- to control the productivity at site
- to correct and instruct when and where necessary
- to be able to arrange for the necessary operational inputs
- to control costs
- to receive sufficient feedback (experience) to prepare further bids.

In order to carry out all these control functions you need to be fully aware of the contents of the contract, and be fully informed at all times about what is happening on site. For routine maintenance contracts it may not be necessary for you to be present on site every day as the operations become routine and are technically not very demanding. However, at the beginning of the contract you may have to spend a couple of days on site until all activities are running smoothly. Afterwards it is recommended that you visit the road at least once a week.

This will give you the opportunity:

- to collect the reports from your gangleaders
- to check the quality of work
- to discuss with the gangleaders and the labourers any problems that may have arisen
- to carry out on-the-job training where necessary
- to provide the gangleader with the instructions and work plan for the following week.

Your personal involvement in supervising and monitoring is extremely important. Only if you are fully in control of all operations at site you will be in a position to avoid losses. It is therefore advisable that you carefully analyse the reports you are getting from the field and compare them with the targets. Where the differences are minor you may not need to worry, but where the difference between the set targets (your plan) and the actual achieved output is great, you need to analyse the cause of this difference. Always check first whether the target was set correctly; maybe your assumptions were unrealistic. If the target seems to be correct you then have to investigate in the field what the cause of the difference was. The gangleader should be in a position to explain.

# *Part II: Routine Maintenance Activities*

## Introduction

The routine maintenance operation can be divided into single maintenance activities, as we have already discussed in Chapter 6 and in previous sections of this chapter. In this section we will now discuss how these activities can be carried out on site. Use the information as a guideline only, as you need to adjust your work method to the particular conditions of the road you are going to maintain. However, it will be necessary for you to know exactly what each and every activity means in terms of the operations, resources and productivity-rate involved. Only if you have a clear picture will you be in a position to estimate the costs and to plan the operations.

The activities presented here may not correspond exactly with the activities in your contract. Sometimes activities are lumped together (e.g. inspection and clearing road from obstructions and debris) or they are broken down into more activities (e.g. 'clean culvert' becomes: 1. clean culvert inlets, 2. clean culvert pipe, 3. clean culvert outlets).

It is also possible that you might get a contract which covers only one or two activities (e.g. the opening of all culverts, mitre drains and side drains). Sometimes this is necessary at the beginning of a maintenance period to clear the existing maintenance backlog. When these activities are completed a normal routine maintenance contract will be issued.

The activities we will discuss are as follows:

1. Inspection and removal of obstructions and debris
2. Clean culverts and their inlets and outlets
3. Clean/repair drifts/splashes and their inlets and outlets
4. Repair culvert headwalls and aprons
5. Clean side drains and excavate to original size
6. Repair/construct scour checks
7. Clean/repair mitre drains and excavate to original size
8. Fill pot-holes and ruts in carriageway
9. Repair shoulder and slope erosion
10. Grub edge and reshape carriageway
11. Cut grass
12. Clear bush.

# Activity 1: Inspection and Removal of Obstructions and Debris

## DESCRIPTION/REMARKS

The activity includes an inspection of the complete length of road under contract. The inspection has to be carried out by the gangleader(s) on a daily basis. Any small obstructions can be removed immediately when detected by the gangleader. The gangleader is supposed to record any work that needs to be done immediately, e.g. blocked culvert during rainy season. He or she also records any larger emergency work needed and immediately informs the contractor, who in turn consults the client.

## WORK METHOD

1. Check road on every working day for emergency maintenance requirements.
2. After every rainfall check whether inlets and outlets of all culverts, drifts or splashes are blocked.
3. After every rainfall check whether all side drains, mitre drains and catchment drains are blocked.
4. After every rainfall check whether obstructions on the carriageway are blocking the road.
5. If any obstructions or debris have to be removed along the section decide how many labourers are required to carry out the work.
6. Organize the labourers and issue task rates.

## REQUIRED HAND TOOLS

According to type of work.

# Activity 2: Clean Culverts including Inlets and Outlets

## DESCRIPTION/REMARKS

The activity includes the removal of all silt and debris from inside the culvert and the area of the headwalls. The inlet and outlet ditches must also be cleared of vegetation, silt and debris. If the ditches are not silting or eroding, and they are to the current depth and profile, then the grass should be cut leaving the roots to bind the surface together.

If rainwater ponds in the culvert, inlet or outlet, the drains should be checked for the correct gradient. If the gradient is below 2% then the drain has to be set out again by the gang-leader using a line level, ranging rods and/or boning rods.

## WORK METHOD

1. Remove silt and debris from inlet, culvert and outlet. Set out outlet width and length, using pegs and strings if necessary.

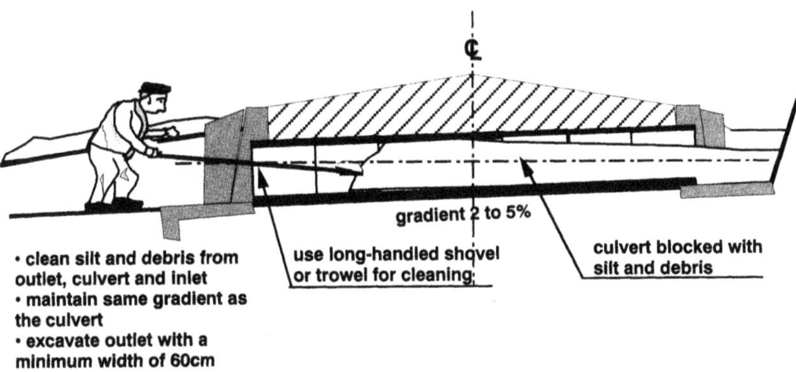


Figure 7.5 *Culvert cleaning*

2. Check gradient of outlet if culvert frequently silts up and re-establish outlet gradient if necessary.
3. If gradient cannot be improved, inform the client that the culvert would need raising.

## REQUIRED HAND TOOLS

- shovel
- hoe
- long-handled shovel and trowel
- gangleader's measuring aids
- line level, ranging rods and/or boning rods
- pegs and strings.

# Activity 3: Clean/Repair Drifts/Splashes including Inlets and Outlets

## DESCRIPTION/REMARKS

Drifts and/or splashes are constructed instead of culverts where the natural gradient of the area does not allow the installation of culverts. Drifts and splashes are usually very delicate when it comes to establishing the minimal gradients; accurate work is essential.

The activity includes checking the gradients and removing all silt and debris. The inlet and outlet drains must also be cleared of vegetation, silt and debris. If the drains are not silting or eroding, and they are to the current depth and profile, then the grass should be cut leaving the roots to bind the surface together.

If rainwater ponds in the drift/splash bed, inlet or outlet, the drains should be checked for the correct and *uniform* gradient. If the gradient is 0% or not uniform the drain has to be set out again by the Gangleader using a line level, ranging rods and/or boning rods.

## WORK METHOD

1. Check all levels in drift/splash bed, inlet and outlet.
2. Establish the correct level of the drift bed.
3. Correct the wrong bed levels:
   - Always check the levels of inlet and outlet first. If the levels of inlet and outlet are correct, move to the next point.
   - Re-establish the correct levels as described under points 1 and 2.
   - Excavate/fill the drift bed.
   - Provide the drift bed with a new gravel layer.

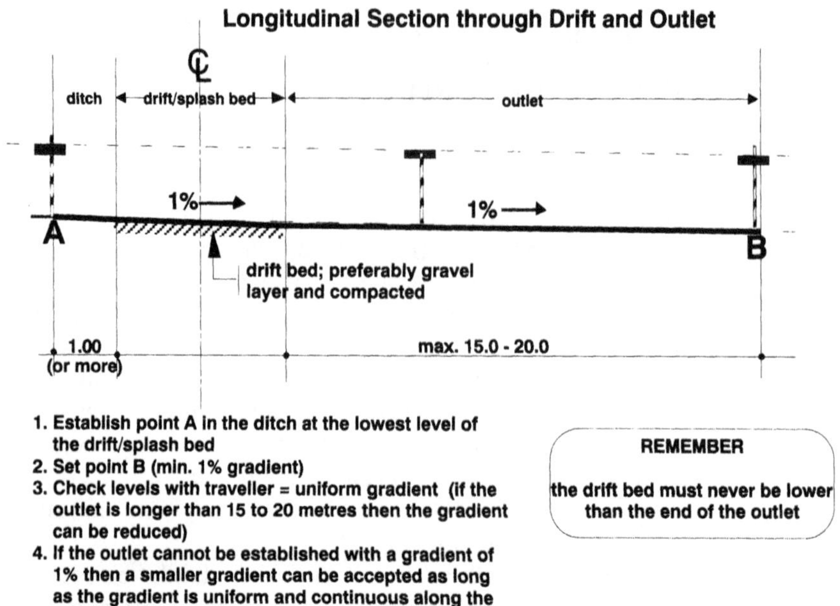


Figure 7.6 *Drift/splash*

4. Remove silt from bed:
   - Remove silt from the drift bed down to the firm surface.
   - Check the levels again.
   - Also check the levels of the outlet. If necessary, correct the outlet.
   - If the original surface under the silt is poor then follow the procedure described under point 6.

Figure 7.7 *Remove silt from drift/splash bed*

5. Repair erosion in bed:
   - Remove poor material and excavate a uniform base.
   - Bring in good material (preferably gravel) in layers.
   - Compact layers of 10cm maximum with earth rammer.
   - Establish a smooth correct level with 1% gradient to the outlet.
6. Remove deformation and/or poor material from bed:
   - Proceed in the same way as for erosion repair, point 5.
7. Monitor the work progress. If necessary demonstrate how to carry out the work, check the work standard and approve the achieved task when the job is finished.

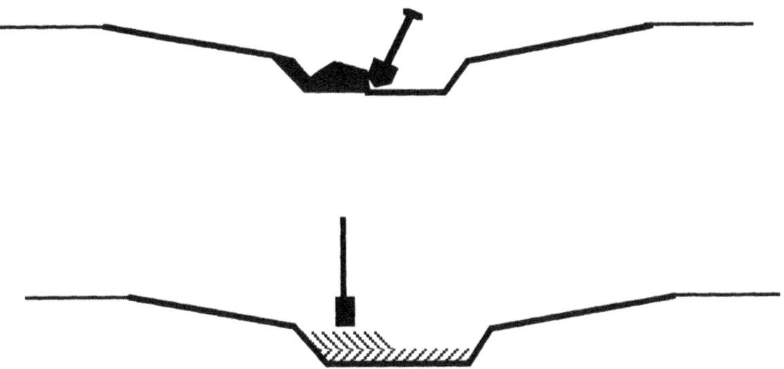

Figure 7.8 *Repair drift/splash bed*

## REQUIRED HAND TOOLS

- shovel
- hoe
- pickaxe
- 2 wheelbarrows
- 1 earth rammer
- gangleader's measuring aids
- line level, ranging rods and/or boning rods
- pegs and strings.

# Activity 4: Repair Culvert Headwall and Wingwalls

## DESCRIPTION/REMARKS

Many of the culverts have dry-stone headwalls, wingwalls and aprons. Where these are damaged it is necessary to repair them immediately with new stones, or stones brought back into place. Try to have one labourer per gang who is specialized in this sort of work.

Repairing mortared masonry work is not usually a routine maintenance activity. However, should this be included in a contract, it is necessary to arrange for a qualified mason to carry out such repair work.

## WORK METHOD

1. Inspect headwalls, wingwalls and aprons regularly, but especially after rainfalls.
2. Replace dislodged stones or fill in new ones. If necessary, shape them to fit properly into the open gaps.
3. Backfill where necessary with gravel or good quality soil.

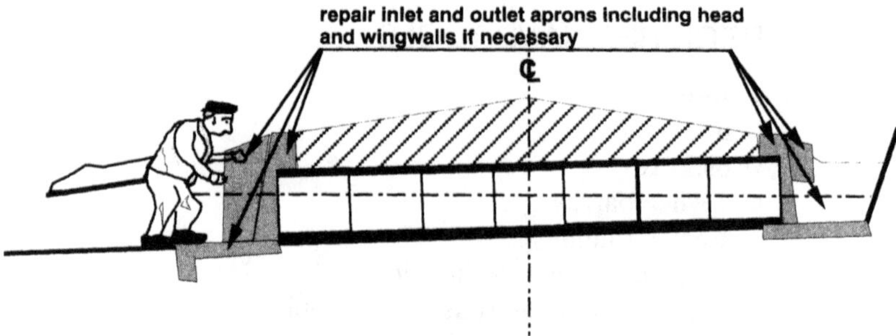


Figure 7.9 *Repair headwall, wingwalls and aprons*

## REQUIRED HAND TOOLS

- shovel
- hoe
- pickaxe
- mason's hammer
- masonry tools for mortared work
- wheelbarrow.

## MATERIAL

- additional stones
- for mortared work: cement and sand.

# Activity 5: Clean Side Drains and Excavation to Original Size

## DESCRIPTION/REMARKS

Where the side drain is established to the correct depth and profile with a grass cover and no erosion, only grass cutting is required.

Where the drain has silted more than 10cm in depth, the vegetation and silt must be removed to the correct depth and profile (use ditch template to check) – this does not apply at the location of scour checks where silting is normal.

All debris and other material from the side drain must be removed well clear of the road and drainage system to prevent it being washed back.

## WORK METHOD

1. Using the ditch-slope template and a spirit-level, a 50cm wide slot is excavated to the correct ditch profile every 10m. This acts as a guide for the labourer to excavate the ditch and slope.

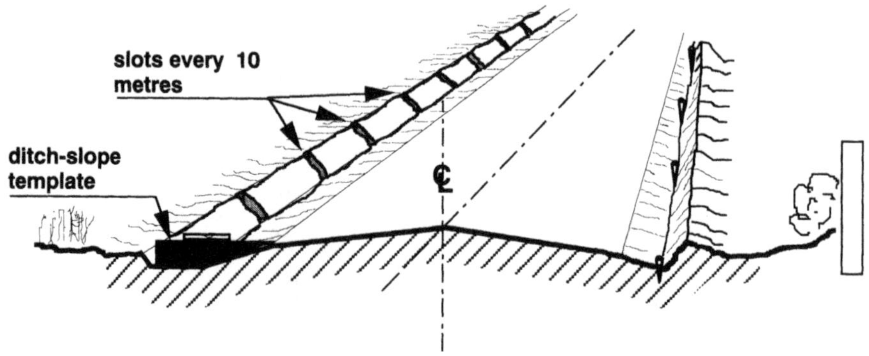


Figure 7.10 *Establish side drain profile (V-ditch cut by grader, form new V-ditch suitable for labour-based methods)*

2. Remove all material from the side drain well clear of the road.
3. Re-check correct side drain levels and profile.

REQUIRED HAND TOOLS

- ○ shovel
- ○ hoe
- ○ ditch-slope template and spirit-level
- ○ pegs and strings.

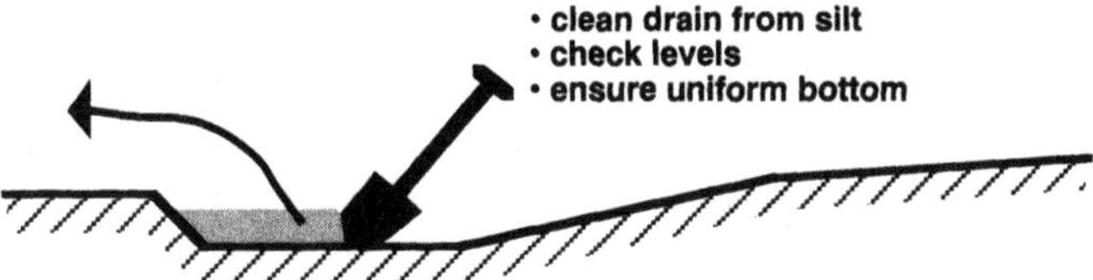


Figure 7.11 *Clean side drain*

# Activity 6: Repair/Construct Scour Checks

## DESCRIPTION/REMARKS

Where drain gradients are steeper than about 4% the water flows at high speed. If no protective measures are taken, scouring is likely to occur on erodible soils. The simplest way of dealing with scouring is by reducing the volume of water (mitre drains at frequent intervals). In addition, replacement or new scour checks should be constructed to reduce the speed of the water. They hold back the silt carried by the water-flow and provide a series of stretches with gentle gradients interrupted by small 'waterfalls'.

A detailed description of scour checks and their construction is provided in Chapter 5, Drainage and Erosion Control.

## WORK METHOD

1. Identify road sections where the ditch gradient is more than 4% using a line-level.
2. Identify the precise ditch gradient and space scour checks according to the gradient (see Figure 7.12).
3. Cut pegs (min. 50cm long) and/or prepare stones.
4. Construct scour check with the correct profile. Use the scour check template for control.
5. Construct a stone apron below the scour check of minimum length 40cm. Dig stones into the ground.

## REQUIRED HAND TOOLS

- template + spirit-level
- sledge-hammer and mason's hammer
- bush knife and shovel

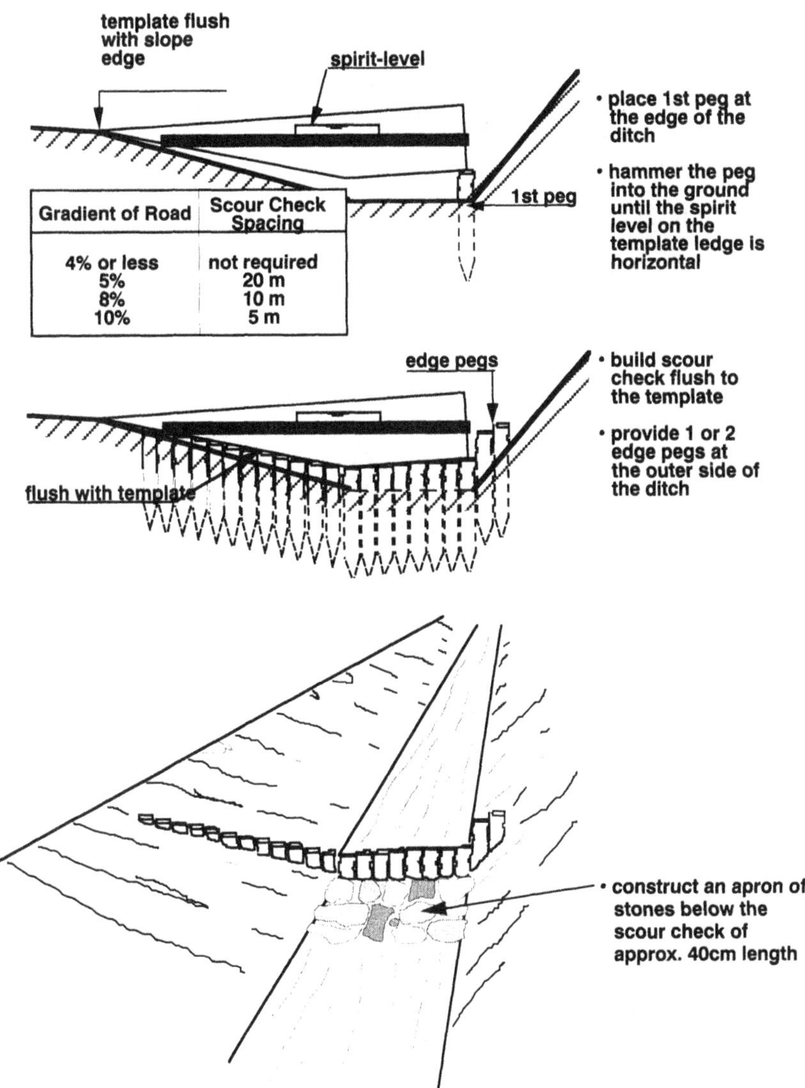


| Gradient of Road | Scour Check Spacing |
|---|---|
| 4% or less | not required |
| 5% | 20 m |
| 8% | 10 m |
| 10% | 5 m |

Figure 7.12 *Establish a scour check*

# Activity 7: Clean/Repair Mitre Drains and Excavate to Original Size

## DESCRIPTION/REMARKS

Mitre drains have to be clear of silt, vegetation and debris. If mitre drains are silting check the gradient, and correct if necessary. Mitre drains should have a minimum gradient of 2%, but not less than the side drain.

If mitre drains are eroding start by checking the gradient and correct as necessary. In addition scour checks can be built in the mitre drain or the number of mitre drains can be increased to reduce the volume of water. Before starting major works such as building new mitre drains always consult the client and obtain written confirmation where your extra payment is indicated.

If mitre drains are not silting or eroding, and they are to the correct depth and profile, then the grass should be cut leaving the roots to bind the surface together.

A precise description of scour checks and their construction is provided in Chapter 5, Drainage and Erosion Control.

## WORK METHOD

1. Identify those mitre drains that i.) silt, ii.) erode, iii.) drain properly.
2. Check the correct gradient.
3. Remove silt, vegetation and debris and dispose of all material well away from the drain.
4. Shape the mitre drains according to the standard size (set out using pegs and strings).
5. Reinstate the drain block.
6. Repair eroded mitre drains by installing scour checks or adding further mitre drains to reduce the water volume.

## REQUIRED HAND TOOLS

- hoe and shovel
- strings and pegs
- ranging rods, boning rods and/or line level

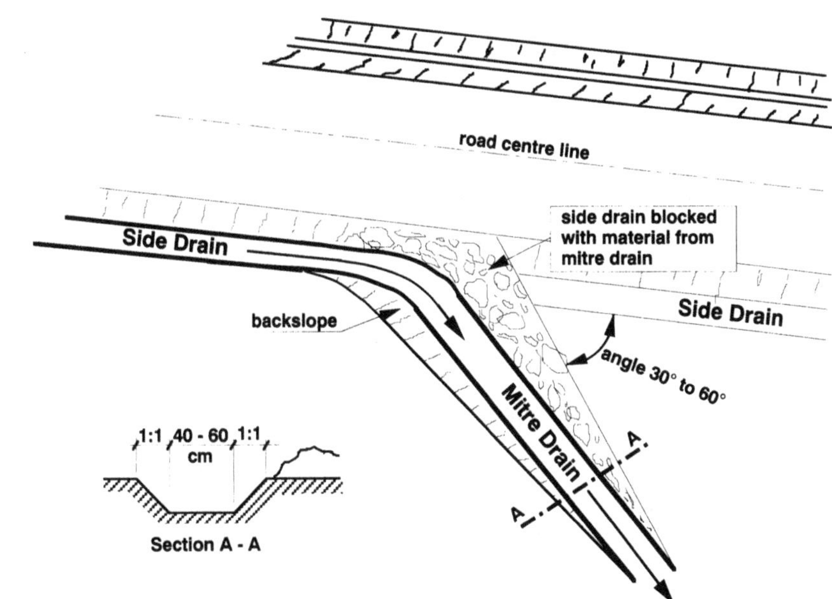


Figure 7.13 *Mitre drain*

# Activity 8: Fill Potholes and Ruts in Carriageway

## DESCRIPTION/REMARKS

Where pot-holes and ruts occur in the carriageway they should be filled with gravel, if available. Otherwise small stones and good soil should be mixed and used, but not topsoil or black cotton soil.

## WORK METHOD

1. Remove weak or soaked material from the pot-hole or rut and dig out until firm ground/material is reached.
2. Fill with new material in layers of not more than 10cm.
3. Compact each layer using an earth rammer.
4. The last layer should be slightly higher than the existing surface (to allow for settling).

## REQUIRED HANDTOOLS

- pickaxe
- hoe and shovel
- wheelbarrow
- earth rammer

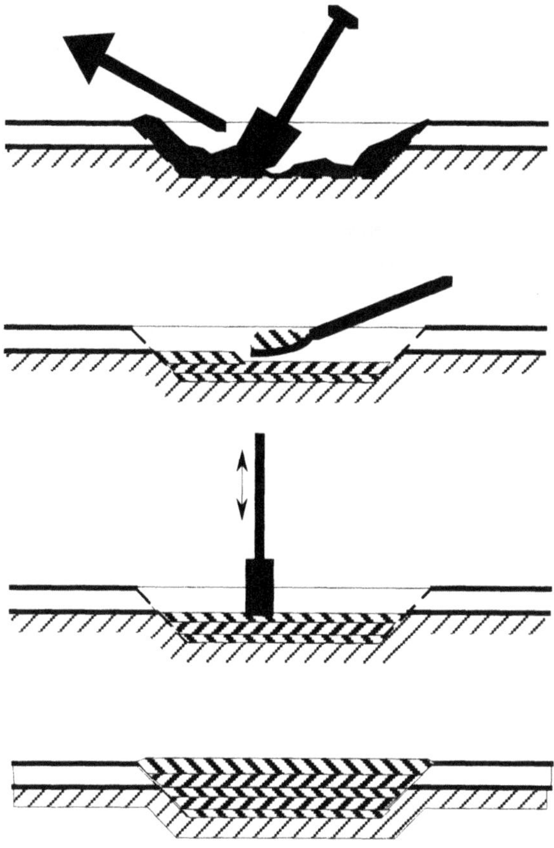

Figure 7.14 *Fill pot-holes/ruts*

# Activity 9: Repair Shoulder and Slope Erosion

## DESCRIPTION/REMARKS

Erosion gullies can be formed by water running over the edge of the carriageway and road shoulder. The gullies should be repaired and filled with gravel where available, or soil. Although no vegetation should be allowed on the carriageway, grass should be established on the sloping shoulders to resist erosion forces. Usually suitable grasses may be found at the site. They should be replanted in the shoulders.

## WORK METHOD

1. Fill the gullies with gravel or good soil.
2. Compact the filled area with an earth rammer.
3. Plant grass along the shoulder.

## REQUIRED HAND TOOLS

- hoe and shovel
- rake
- wheelbarrow
- earth rammer
- water container.

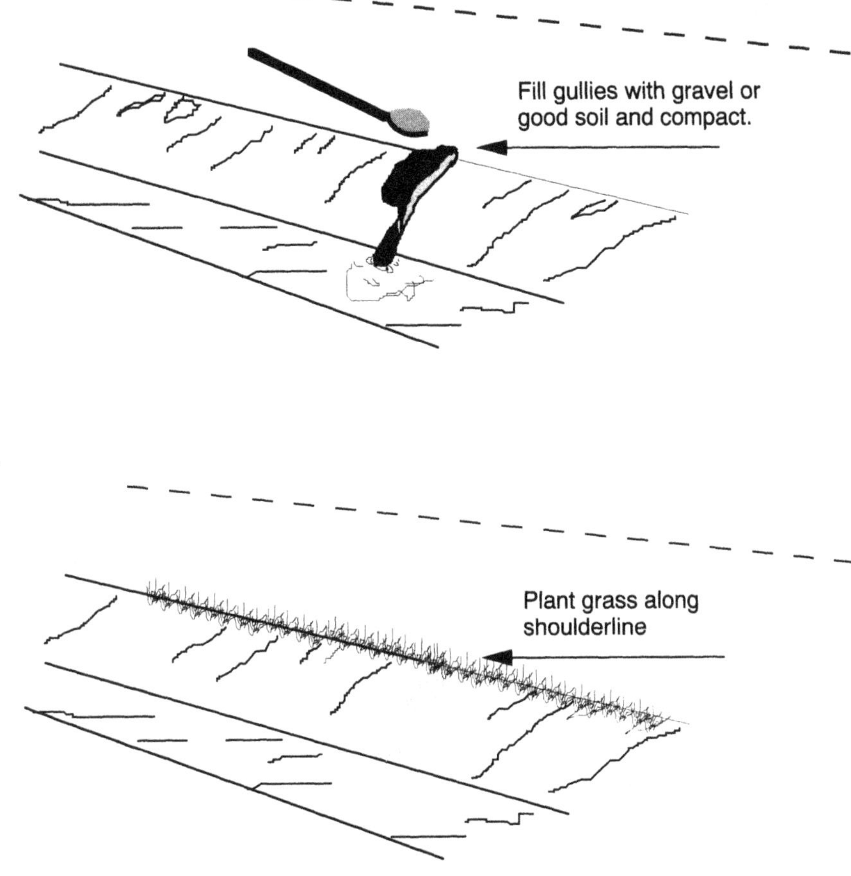


Figure 7.15 *Repair erosion on shoulder*

# Activity 10: Grub Edge and Reshape Carriageway

## DESCRIPTION/REMARKS

Grass growing on the edge of the carriageway can prevent rainwater causing erosion to the shoulder, but on the other hand it may also grow into the carriageway and hinder rainwater from draining easily.

The grass roots should be removed from the carriageway. A stringline should be set out at the edge of the carriageway (beginning of shoulder) as a guideline for the activity. Debris should be removed well clear of the road and the drainage system.

In time, the camber of the carriageway will be reduced due to traffic and weather. The camber should be reshaped by bringing back material from the edge towards the centre line. A camber board or a straight edge should be used to control the uniformity and gradient of the reshaped carriageway area.

## WORK METHOD

1. Set out the shoulder–carriageway line using pegs and strings.
2. Grub the edge using a hoe towards the slope and remove the material (grass and roots) well away from the road and the drains.
3. Cut the shoulder and/or road edge back to the original shape.
4. Fill material into the ruts using rakes or shovels.
5. Check the camber using the camber boards.

## REQUIRED HAND TOOLS

- hoe and shovel
- rake
- camber board or straight edge with spirit-level.

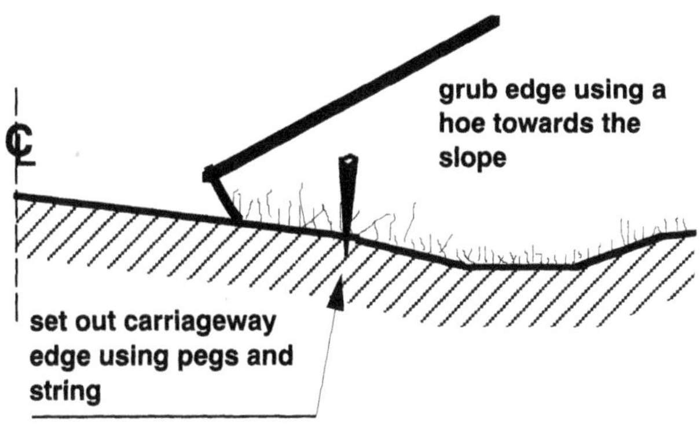


Figure 7.16 *Grub carriageway edge*

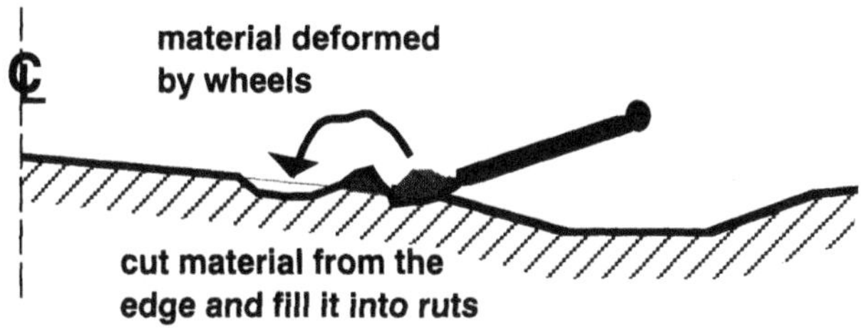


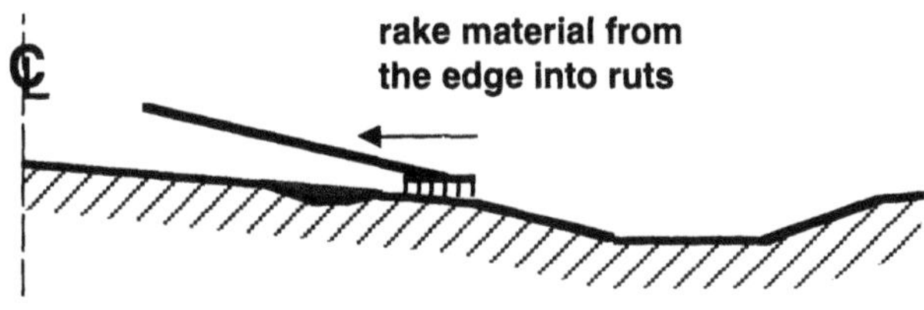


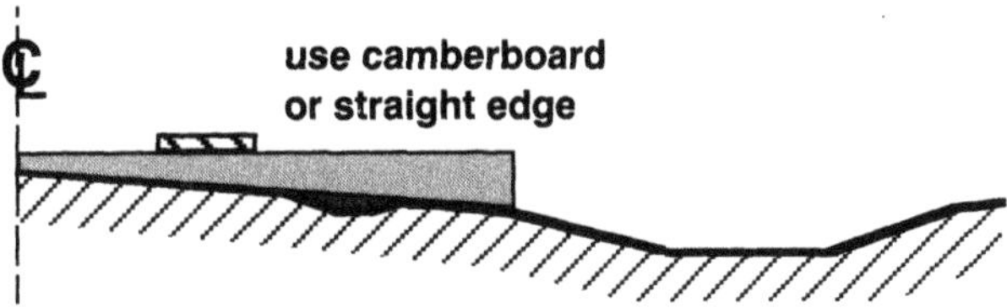


Figure 7.17 *Reshape carriageway*

# Activity 11: Cut Grass

## DESCRIPTION/REMARKS

Grass on shoulders, in side drains and mitre drains should be kept short (about 5cm high).

## WORK METHOD

1. Identify width of grass that should be kept.
2. Cut grass short (about 5cm high).
3. Remove cut grass well away from the road.

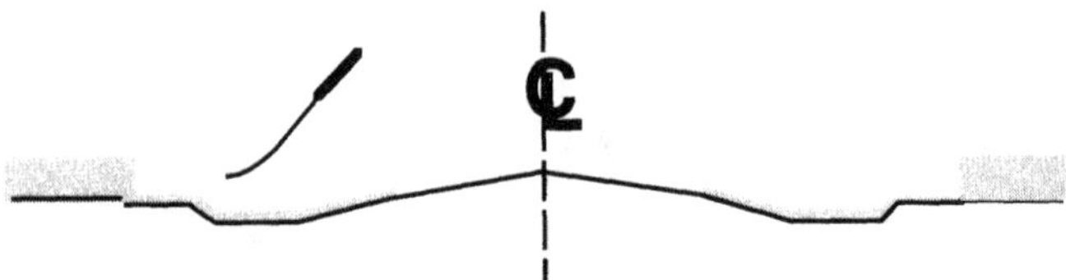

Figure 7.18 *Cut grass on shoulders and drains*

## REQUIRED HAND TOOLS

- grass slasher
- rake.

# Activity 12: Clear Bush

## DESCRIPTION/REMARKS

All bushes and branches overhanging the road, shoulders, side drains, mitre drains and traffic signs should be cut down and disposed of safely. The debris can be burnt if there is no risk to traffic or surrounding crops/vegetation. Otherwise it should be removed well behind the side drain.

## WORK METHOD

1. Cut the bushes.
2. Remove the debris well clear of the road.

Figure 7.19 *Cut bush*

## REQUIRED HAND TOOLS

- bush knife
- axe
- rake.

# CHAPTER 8: REGRAVELLING

## Learning Objectives

After you have completed this chapter you should know:

- the required operational approach to regravelling and how to organize work at site
- how to establish and manage a regravelling site
- how to plan and monitor regravelling
- how to carry out all regravelling activities.

## Introduction

This chapter discusses regravelling as a maintenance operation. Regravelling is basically the same operation as the initial gravelling and we will therefore use the term 'gravelling' in the text.

In order to benefit fully from this chapter you need to understand the contents of the preceding chapters:

- Chapter 3; Soil Mechanics
- Chapter 4; Equipment and Tools
- Chapter 5; Introduction to Labour-Based Road Construction
- Chapter 6; Road Maintenance

If you are unsure whether you have fully understood these previous chapters, you should go back and repeat the necessary sections before starting this chapter.

In chapter 6 you had a chance to learn about the general management aspects and the principles of road maintenance. In this chapter we will take a close look at your tasks as a contractor undertaking regravelling work.

The client (road maintenance manager) has identified the need for regravelling and made sure that funds are available for your contract. The contract tells you how the client wants the job to be carried out. The contractor is then fully responsible for achieving the standards specified, within the time limits set, and at the quoted cost.

## The Gravel Layer

People generally want to be able to use the roads in both wet and dry weather. Roads also have to take heavy vehicles like trucks and buses without being damaged. These requirements, which are taken into consideration at the design stage, of course also influence the maintenance stage. The gravel surface deteriorates with time due to the effects of traffic and weather.

A big problem is that the vehicle load imposed on the road is concentrated. Compare how easy it is to punch a pencil into soil, while you could not make a dent by pushing hard with the flat of your hand. The road builder puts a surface material between the tyre and the natural soil to spread the load, i.e. the soil is not deformed. Figure 8.1 shows how the gravel layer takes the concentrated traffic load and spreads it over a larger area of natural soil.

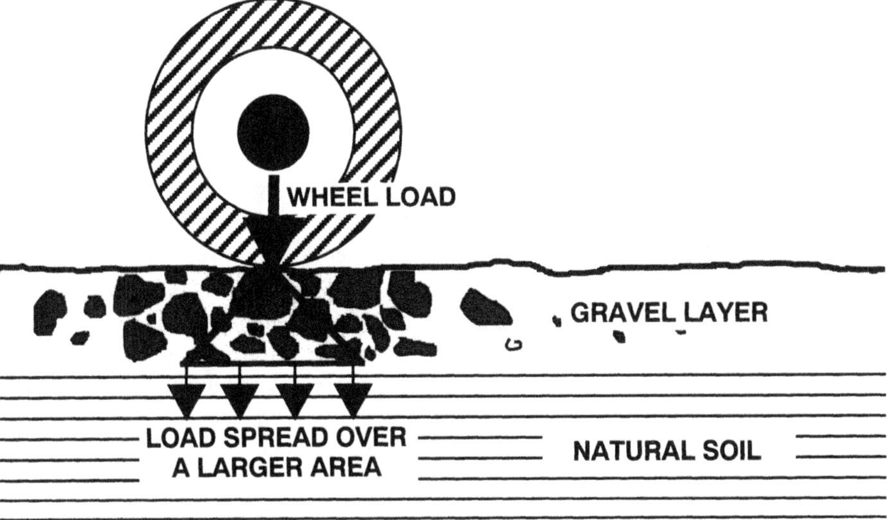


Figure 8.1 *Load spreading by surface layer*

More information on the surface layer, including the required qualities, is presented in Chapter 3, Soils.

As you know, the quality and quantity of the work is specified by the client. For gravel roads there are generally national standards, such as design guidelines defining, for example, thickness and quality of the gravel layer. In most cases the required compacted thickness of gravel is a minimum of 12 to 15cm. The width can differ very much and depends on the standard cross-section of the particular road. The quality of the gravel is specified in your contract. In addition, the client will often designate a suitable quarry to use, but remember, the contractor is responsible for what is put on the road. You as a contractor therefore have to ensure that the quality of gravel from the quarry corresponds with the requirements set out in the contract.

The gravelling operation involves a substantial amount of money. A major mistake here, resulting in having to redo work can put your company at risk. The client also considers it an expensive operation, so they will always put high emphasis on adherence to the specifications. They want to get good value for their money. As this operation can determine whether your business is run at a profit or a loss, you have to look out for possible sources of problems, such as a change in the quality of gravel coming from the quarry. However, most important is to put a lot of effort into carefully planning the implementation of a project.

# Organization and Planning of Works

## GENERAL

In terms of organization, labour-based gravelling is a demanding job. A site manager has to find the best balance between a large number of labourers and pieces of hauling equipment. The speed of the operation is governed by the output capacity of the equipment, so the labour force has to be adjusted to match that capacity. In other words, you need labourers to excavate, load, off-load and spread as much gravel as the hauling equipment can transport in a day. In addition you might need to take into account a change in the hauling distance and the risk of equipment breakdown.

It is easy to understand that the management of a gravelling site is difficult and that there are many opportunities to lose money. But on the other hand, with good site organization, proper planning and supervision it is possible to make a substantial profit. What is absolutely essential is to have experienced and capable everyday site supervision. Do not take any chances with site management of regravelling works. Either manage the job yourself or put a very experienced and skilful foreman to look after it. A lot of money is at stake, and failure or success on a larger regravelling contract has often proved to be the difference between a thriving small-scale contractor and one who has gone bankrupt.

## SITE INSPECTION

Before you start preparing a tender and work plans you must visit the site and carry out a thorough site inspection. Make sure you collect all the data necessary to prepare a realistic tender. To help you, we propose the following checklist:

*Quarry*

- How much gravel is available in the particular quarry? You should assess the volume of gravel that you can get from the quarry and compare it with the gravel volume you need for the road or section of road to be gravelled from this quarry. Try to estimate the thickness and area of the gravel deposit. You will probably need to dig some test holes and carry out tests in a systematic manner.
- Is the quarry located in low-lying terrain? Low-lying quarries may well become unworkable when it rains.
- Does the quality of the gravel match the contract specifications? (see test procedures in Chapter 3).
- Is water available at or near the quarry site?
- How much of the overburden will have to be removed to reach the gravel material, and where can it be deposited (haul distance from quarry to depot)?
- Is the land owned by the client? Will there be a charge for its use? If it is owned by a local landowner, how much compensation will have to be paid? There are usually established procedures for compensation, and standard prices. You can normally obtain advice and data from the client, as you both have an interest in getting good material at a reasonable cost.
- Is there a danger of soil erosion when you exploit the quarry, and what would be the required protection work? – for example filling the quarry with overburden at the end of the job, etc.

*Quarry Access*

- What is the length of the access road that you need to construct or improve, and through what type of terrain does it run?
- How much work needs to be done to improve the access road so that the hauling can be carried out undisturbed? (Try to estimate the required work input and the relevant costs.)

*Hauling*

- What is the average hauling distance for this job (there may be more than one quarry)? Definition: The average hauling distance is the distance from the quarry to the middle of the closest and furthest road sections that you need to gravel.
- What is the condition of the haul route when improved (gradient, roughness, river crossings, soil, etc.)?

- Can your equipment be used for hauling on this particular haul route?

*Compaction*

- Where can water be collected for compaction?
- What is the condition of the access road to this place and what improvement work is required?

*Camp Site and Labour*

- Where are the best locations for site camps (considering distance to site, water availability, access to market and health centres, security, etc.)?
- How many camp sites are required at any one time? – do you need to shift camp sites during the work? – how many times?
- Is any preparation work required for the camp, e.g. access road, bush clearing, etc.?
- What is the average walking distance for the labourers, from their homes to the camp, and from the camp to the sites (gravelling site, quarry, reshaping site)?
- What is the availability of skilled/unskilled labourers?
- Can you get support from the local authorities to recruit and organize labour?
- Who are the influential people in the area of your operation?
- What is the transport distance from the company's base to the camp?
- What mode of transport is required and how many trips (camp facilities, tools, fuel, spares, material, supervision, payments, etc.)?

*Road Condition*

- What is the existing condition of the road before regravelling work commences, in comparison to the required standards?
  - drainage (open or blocked? sufficient mitre drains ? culverts in place? drainage gradients correct? scour checks in place?, etc.)
  - camber (is the gradient sufficient or does it need to be improved?)
  - roughness of carriageway (is heavy reshaping required?)
  - structures (are the structures in good condition or do they require repair work?)
  - is there any gravel left on the carriageway? if so, how much?
  - where are the trouble-spots on the road and of what nature are they?

– is there soil erosion along the road caused by water running of the road, and what must be done to avoid it?

It is important that you have a clear picture of the situation of the road you have to gravel, the quarry or quarries that you need to haul the gravel from, including their potential gravel volume, and the quarry access roads. It is sometimes useful to prepare a sketch especially when you have to operate from more than one quarry. This can be a simple strip map with the most important data included (for an example, see Figure 8.2).

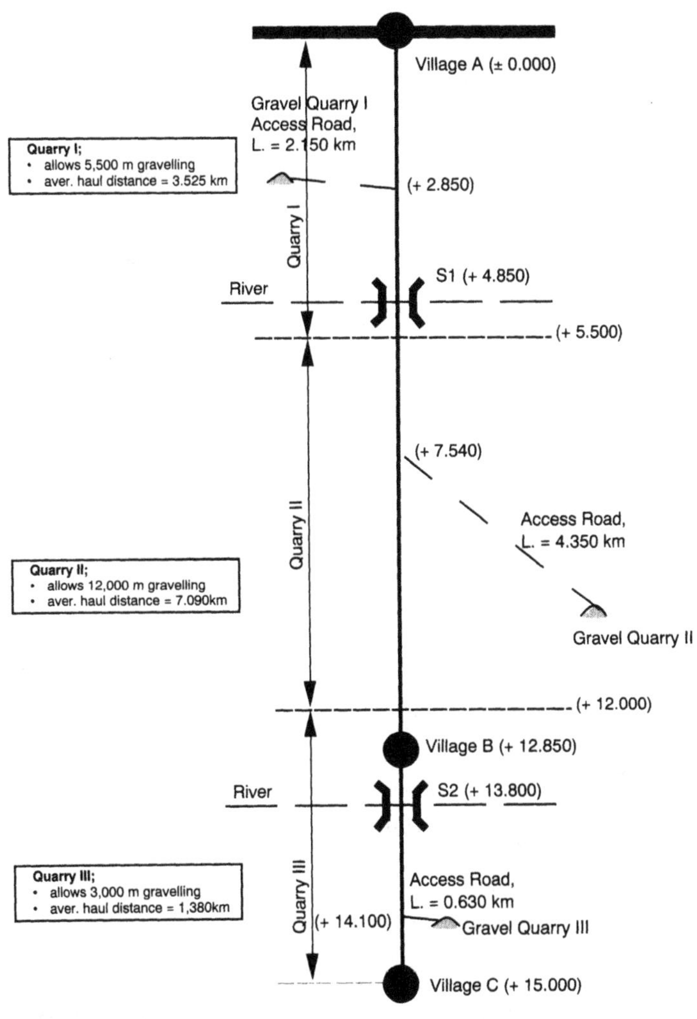


Figure 8.2 *Road strip map with quarry locations*

## OPERATIONAL APPROACHES

The gravelling works are divided into two main activities:

1. Preparation, which consists of:

- reshaping the road
- preparing the quarry access road
- preparing the quarry
- excavation and stockpiling of gravel.

2. The gravelling operation, which consists of:

- excavation of gravel in the quarry (continuation)
- loading
- hauling
- off-loading and spreading
- compaction
- stockpiling of gravel for maintenance.

All of these activities except hauling are carried out with labour-based methods. How you organize the gravelling activities is determined by the hauling equipment you have available. In Chapter 4, Equipment and Tools, the requirements in terms of equipment for gravelling works is discussed, resulting in the presentation of two principal alternatives in terms of hauling equipment used:

A. Tractor-trailer combinations normally used for hauling distances of up to 10km.

B. Lorries (tippers or flat-bed) usually most economical for hauling distances over 10km.

As a contractor, you must always identify the possible options you have and make a cost comparison to identify the cheapest alternative. If you own this kind of equipment, a cost comparison is likely to result in favour of using that equipment, with a normal hauling distance. As regravelling contracts tend to be one-off operations, owning equipment might not be the best idea, especially in the beginning when you enter the market. You should always check whether hiring hauling equipment, i.e. paying only for the equipment when it is needed, is a more economical approach.

Whatever option you choose, the main aim is always to make the best use of your equipment when on site. That can only be achieved if you plan your activities well. When it comes to a fairly 'equipment-intensive' activity, such as regravelling, your plans have to be guided by the equipment operation, i.e. by the capacity of the equipment.

It is clear that in regravelling operations the choice of equipment is very important. On the following pages the most important pieces of equipment are reviewed and their advantages and disadvantages are listed. This section gives guidance to help you to choose the best alternative, i.e. what is most suitable for your company. In addition we also look at the potential for animal-drawn hauling, as this may be a viable alternative in some areas.

## TRACTOR–TRAILER HAULING

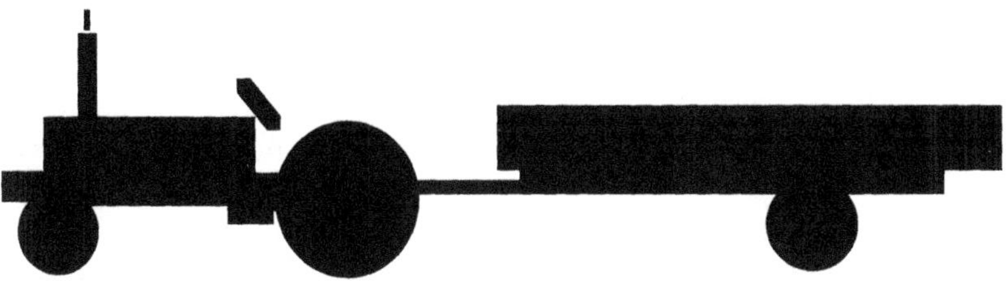

Figure 8.3 *Tractor trailer combinations*

System:
Tractor and trailer combinations are best suited for hauling distances up to 10km. Each tractor should have two trailers so that the tractor can be optimally utilized, e.g. one trailer is loaded in the quarry while the other is hauled to site. For an average regravelling site (e.g. 10km regravelling) 3 – 4 tractors and 6 – 8 trailers will be sufficient to complete the work in a reasonable time (60 to 100m per day). Tractors for this sort of operation should be in the power range of 50 to 60hp while trailers are recommended to have a loading capacity of $3m^3$.

The advantage of using tractors and trailers is their flexibility. The tractor is a power source which can tow many other different types of attachments. Tractors are also widely available in the agricultural sector and are therefore relatively easy to hire. Another advantage is the relatively good back-up service that is available for agricultural tractors; spare parts are available and mechanics are trained to service and repair them.

The disadvantage is that tractors and trailers often become uneconomical when the hauling distance is longer than 10km, as they are too slow.

## LORRY HAULING

Figure 8.4 *Lorry hauling*

System:
For longer hauling distances (usually more than 10km) flat-bed lorries or tippers are often the best alternative. For a typical gravelling site of 10km length, and with 10 to 20km hauling distance, you probably need 4 or 5 lorries to achieve an average output of 60 to 100m per day. Flat-bed lorries are loaded and off-loaded manually.

The major advantage of using lorries is that long hauling distances can be economically managed. In addition, many building contractors already have one or two lorries available in their company, and additional lorries can often be hired from other firms, i.e. you do not need to invest in new equipment.

The disadvantage of using lorries is that they have to stand idle in the quarry while they are loaded by hand and that the gravel needs to be lifted higher from the ground to the loading deck than is the case with trailers. However, if the quarry works, including the loading activity, are properly organized, a good output can be achieved also when labour is used to load and off-load.

## ANIMAL-DRAWN CARTS

Figure 8.5 *Animal-drawn hauling*

System:
Animal-drawn carts, mainly donkey-carts, have been used successfully in some projects where gravel is very close to the road

(maximum 2km hauling distance) and where animal-drawn operations are known to the people of the area. The hauling operation needs to be subcontracted to animal owners. Special attention has to be given to appropriate carts and harnesses.

The advantage of using animal-drawn carts is the high availability and flexibility of this transportation system. You do not need a full fleet of transportation equipment and are not dependent on hiring from other contractors. Utilization of a locally-available resource that provides more employment to people in the area is also an advantage for the country as a whole.

The disadvantage of using animal-drawn carts is their very limited hauling distance. Another problem is that animal owners are often not properly equipped for heavy duty transport like gravel hauling, which requires special carts and high quality harnesses.

## SITE ORGANIZATION

For gravelling it is necessary to subdivide the site into at least three major gangs:

Gang 1: Improvement of quarry access road, preparation of quarry and quarry work
Gang 2: Reshaping/repair of road
Gang 3: Gravelling operations.

The gravelling site has to be under the overall supervision of a foreman, while the different gangs can be supervised by gangleaders. The large gang at the quarry may be further subdivided into smaller gangs according to activities, thus they are easier to supervise by the gangleaders.

It is important that you set up an organization system that specifies the different duties and responsibilities clearly.

## SITE CAMP ESTABLISHMENT

The setting-up of the site camp should follow the guidelines provided in Chapter 5. If the quarry is far away from the road, it is sometimes necessary to set up two camps; one for the quarry gang and another for the reshaping and gravelling gangs.

The difference, compared to a construction camp, is that sufficient accommodation must be provided for the plant operators and that fuel and spare parts have to be kept at site (see guidelines in Chapter 4, Table 4.4).

Where two camps are required, hand tools need to be shared and kept in both camps. The general requirements for a total of 120 labourers are shown in Table 8.1.

**Table 8.1** Recommended principal hand tool requirements

| *Hand tools* | *Quarry Gang* | *Reshaping and Gravelling Gang* |
|---|---|---|
| Shovels | 70 | 30 |
| Hoes | 20 | 30 |
| Mattocks | 50 | 20 |
| Pickaxes | 70 | 10 |
| Bush knives | 10 | 10 |
| Wheelbarrows | 10 | 5 |
| Crow bars | 8 | 2 |
| Spreaders/rakes | 5 | 20 |
| Sledge-hammers | 4 | 2 |
| Tape measures | 2 | 2 |
| Buckets | 10 | 5 |
| Standard measuring aids | | |
| – Ditch templates | – | 2 |
| – Camberboards | – | 2 |
| – Ranging rods | – | 6 |
| – Boning rods | – | 1 set |

Site stores need to be properly managed, as tools and equipment represent a lot of your money. A good way to ensure proper management is to employ a site store-keeper whom you trust fully.

The duties of the site store-keeper should consist of:

- receiving tools, material and spare parts from your main store
- keeping tools, material and spare parts in an orderly manner in the store
- issuing tools, material and spare parts to the labourers
- receiving tools, material and spare parts back from the labourers
- maintaining proper store records
- undertaking tool maintenance work, such as replacement of handles, sharpening of blades, etc.

To facilitate the paperwork linked to these tasks, you should provide your site store-keeper with two essential reporting forms:

A. Site store record

B. Site issue form for tools and equipment

These records will facilitate your keeping track of your equipment. They will make it easier when tracing loss of equipment, theft, and so on, because you will discover it early and can intervene immediately, thus limiting your losses.

*A. Site Store Record*
The site record should register all the items which are kept at the site store. It must indicate when a particular item has been issued (material, fuel, oils, lubricants, spares and other consumable items), to whom it has been issued, when it has been received and what the stock balance is. This record can be a simple booklet which you prepare for the site store keeper and could look like this:

**Table 8.3** Site store record

ITEM: *Pickaxes*

| Date | QUANTITY received | issued | balance | Delivery Note | Received from / Issued to | Signature |
|---|---|---|---|---|---|---|
| 12.5.96 | 80 | | 80 | A.2365 | Main Store | **** |
| 16.5.96 | 80 | 12 | 68 | A.5678 | Main Store for repair | **** |
| | | | | | | |
| | | | | | | |
| | | | | | | |
| | | | | | | |
| | | | | | | |

You need to use a separate page for each item. Items which are used very frequently, such as fuel or certain spare parts, require several pages. This record will also help you when carrying out your inspections on site.

*B. Site Issue Form for Tools and Equipment*
This form is issued to record the tools you issue to the labourers on a daily or weekly basis. Since you might have up to 120 labourers on a site, you can easily imagine that without proper records you would soon lose control over the equipment. You have to teach your site store-keeper to keep the records accurately and immediately report any loss of tools or equipment so that you can recover the lost item from the salary of the labourer.

It is advisable for you to pay your store-keeper well so that he remains loyal to you and your company. A proposed site issue form is shown in Table 8.4.

**Table 8.4** Site issue form for tools and equipment

| EMPL. NO. | DATE OUT | ITEMS | | | | | DATE IN |
|---|---|---|---|---|---|---|---|
| 86 | 12.5. | HOE | PICKAXE | | | | |
| 87 | 12.5. | HOE | PICKAXE | | | | |
| 88 | 12.5. | HOE | SHOVEL | RAKE | PICKAXE | | |
| 90 | 12.5. | HOE | SHOVEL | RAMMER | | | |
| 93 | 12.5. | HOE | | | PICKAXE | | |
| 94 | 12.5. | CROWBAR | PICKAXE | SHOVEL | WHEELBAROW | | |
| | | | | | | | |
| | | | | | | | |
| | | | | | | | |

You can decide to issue tools on either a daily or a weekly basis. When a particular tool is returned, it must immediately be crossed out in the record. Any item not returned by a labourer should be clearly marked and the labourer asked to return it without delay. If the tool is not returned the cost for a new item will be deducted from the salary. It is advisable to fix the prices slightly higher than what the item would cost in the local shops so that your store does not become an attractive self-service store. When the items are returned your record looks like this:

**Table 8.5** Site issue form for tools and equipment

| EMPL. NO. | DATE OUT | ITEMS | | | | | DATE IN |
|---|---|---|---|---|---|---|---|
| 86 | 12.5. | ~~HOE~~ | ~~PICKAXE~~ | | | | 12.5. |
| 87 | 12.5. | ~~HOE~~ | ~~PICKAXE~~ | | | | 12.5. |
| 88 | 12.5. | ~~HOE~~ | ~~SHOVEL~~ | RAKE | ~~PICKAXE~~ | | 12.5. |
| 90 | 12.5. | ~~HOE~~ | ~~SHOVEL~~ | ~~RAMMER~~ | | | 12.5. |
| 93 | 12.5. | ~~HOE~~ | | | ~~PICKAXE~~ | | 12.5. |
| 94 | 12.5. | ~~CROWBAR~~ | ~~PICKAXE~~ | ~~SHOVEL~~ | ~~WHEELBAROW~~ | | 12.5. |
| | | | | | | | |
| | | | | | | | |
| | | | | | | | |

## EMPLOYMENT OF CASUAL LABOUR

In labour-based construction and maintenance activities the client has two objectives:

- to get the work done at a reasonable cost, within the stipulated time period, while meeting quality targets
- to create local employment.

With the second objective in mind, the contract will probably set out specific rules for recruiting labour.

The labourers you require for a gravelling site will be employed for only a limited period, so you will offer them employment as casual labourers. It is usually the objective of labour-based projects to involve the local people in the work as much as possible. The labourers, the gangleaders and probably even some skilled labour required for the gravelling operation, will therefore be recruited from the vicinity of the road. They will walk to the site from their homes each day. When the gravelling operation on a section of road is completed you will have to replace the original labour force, as the walking distance between home and site becomes too long (usually approximately 4km is the maximum). Therefore new casual labour needs to be employed at each work site.

Before starting the hiring process you must be fully aware of the local labour laws. Make sure that the contracts you have with your staff and your labourers abide by all the relevant laws and regulations. To give you a quick overview of what sort of issues are involved, the main points to be considered are listed below. They concern:

- a minimum working age set to avoid exploitation of child labour
- a minimum labour wage ensuring a minimum basic standard to everyone working
- equal payment for women and men when doing the same job
- full and prompt payment of wages
- safety and health regulations to reduce the risk of injury and sickness related to the workplace.
- not using forced labour, i.e. every labourer should offer their services voluntarily
- insurance, premiums paid by you, to cover any accident or injury to your workers.

The countries that are members of the International Labour Organization have, through signatures to agreements, given birth to a number of International Labour Standards covering issues such as basic human rights, employment and conditions of work. These standards take the forms of Conventions, which are binding, and Recommendations, which are to be regarded as policy guidelines. In the Annexe to this book, ILO conventions especially relevant for labour-based road works are listed, with comments on their applicability and practical consequences for you as a contractor.

In your contract with the client there may also be clauses stating that women should have an equal opportunity with men to gain employment. That means that you must have a recruitment process that assures equal opportunity for women to apply, and a selection process that does not discriminate against women. The contract may also have a clause stating that a certain percentage of the workforce should be women (often 25%).

In addition, a maximum employment period for casual labour (often three months) might exist. If that time period is exceeded the law may require you to employ the labourers permanently.

Why is it important to comply with these regulations? Firstly, there are, of course, the humanitarian reasons involving the general dignity and safety of fellow citizens. Secondly, there are also solid economic arguments; if your workers are treated fairly they are likely to develop a strong loyalty to you and your company that is usually followed by a willingness to contribute a little extra to your company's work. If your workers feel that reasonable precautions are taken to protect them, their productivity is very likely to rise. This will contribute to your achieving a good profit. And as a real bottom line, your contract with the client usually has a clause saying that if the contractor does not respect local labour legislation the contract can be terminated and your licence to bid for other works within the ministry will be cancelled.

Labour complaints and labour disputes can become both very irritating and very costly for a contractor. We therefore strongly advise you to get in touch with the local labour office and seek their advice on these matters. If that proves difficult, your contractors' association should also be able to advise and assist you on these matters.

Remember: proper employment and time records will always assist you when it comes to legal discussions or disputes.

Most contracts require that the recruitment of the casual labour force is officially announced at least two weeks in advance, and that the terms and conditions of employment are fully explained. It is advisable to involve the local administration in this process, although you should make sure that you are free to recruit only capable and motivated people. During the announcement meeting, as well as during the recruitment meeting, it is important to explain the following details to everybody applying for a job:

- the function and organization of the contract
- the type of work which the job seekers will be expected to carry out
- the employment duration
- the number of labourers to be recruited
- the need to carry an identity card for the recruitment drive
- the terms of employment (see employment form)

- pay rates, timing and arrangements for payment, including the first payment date
- that women and men are equally eligible and welcome to seek employment
- that if there are more capable and motivated job seekers than vacancies, you have a selection system that is generally thought to be fair. (A good example of a fair system is to arrange a secret ballot).

After the introduction, all job seekers will be invited to step forward and form a queue during the recruitment meeting. If more people than required come forward, you will have to proceed with an unbiased selection process. It is important for you that the process is generally felt to be fair in order to avoid accusations of corruption or favouritism.

Here is an example of a secret ballot system tried in several countries. It provides you with an idea of how to arrange the process, although you might not, due to different local practice, be able to follow it in every detail. When designing your own system, remember that the key to a successful process is objectiveness. The job seekers should not, at any time of the procedure, doubt that it is a fair system:

- make sure the local administration is available to witness the employment process
- collect the identity cards from all job seekers and put them into a container
- ask a local leader to blindly pick one ID card and call for the chosen person
- this person then picks the next card and proceeds to your clerk, who prepares a list with all chosen names and immediately fills in prepared employment forms (see example, Table 8.6)
- it is advisable to let the local administration sign the list when everything is finalized
- it is also recommended that about 25% more labourers are put on a waiting list (there are always some labourers who desert work after the first few days and need to be replaced).

This process can, of course, be modified to take account of special conditions, such as a certain quota of women workers, if specified in the contract. If the proportion of women applying is lower than the percentage stipulated in the contract, you can have separate containers for the ID cards of women and men.

An additional effect of this process is that the risk of hiring 'ghost workers' is reduced. All workers have to be present at the recruitment stage.

**Table 8.6** Example of a casual employment form

# CASUAL EMPLOYMENT FORM

| | |
|---|---|
| **COMPANY NAME:** | |
| **PROJECT:** | |
| **NAME: MR/MRS:** | **I.D. NO:** |
| **EMPLOYMENT NO:** | **DATE:** |

1. You are hereby offered employment with casual condition as a:

   with effect from:

2. The terms and conditions of employment are as follows:
   a.) You will be paid ..........NU per day for an eight hour working day or an equivalent task rate. You will not be paid for public holidays or any day not worked regardless of the reason i.e. heavy rains, sickness, etc.
   b.) You will be paid at the end of each month for the days worked.
   c.) You are not entitled to annual leave, housing, transport or any allowances
   d.) Your employment will be terminated by any of the following:
   - at the end of the ..... months period starting on the date mentioned above,
   - at the completion of the project for which you were engaged,
   - at any time at the discretion of the contractor
   - when you absent yourself from work without permission
   - when you do not follow instructions from you supervisors
   - when you behave in a disorderly manner.

   e.) You are responsible for any loss or damage of tools issued to you and the cost of such will be deducted from your pay.

3. By signing the acceptance you indicate that you agree with the terms and conditions of employment set out above.

| | |
|---|---|
| **CONTRACTOR:** | **DATE:** |

I have read / had read to me, and I understand the above terms.
I hereby accept the casual employment on the terms and conditions of this form

**SIGNATURE OF EMPLOYEE:**

**FORM TO BE COMPLETED IN DUPLICATE: ONE COPY FOR THE CONTRACTOR**
**ONE COPY FOR THE EMPLOYEE**

After the selection each labourer should be issued with a personal employment form (retain a copy for your reference). It is important that you keep accurate control of the employment; do not delegate this job to anyone else. Make sure that the names on the employment list are the same as on the muster-roll.

As gravelling works are organized in a sequenced way, it will not be possible to call all the labourers as from the first day of operations. They will have to be called according to the planned work sequence. For example, for the quarry preparation works you might need 60 labourers but later, when you go into full gravelling operations, you will require 120.

*Payment of Wages*

It is important to pay a wage high enough to attract labourers, yet related to the value of the work carried out by the labourer. Before you decide on how much to pay, make sure you are familiar with local laws and what they say about minimum labour wage rates and so on.

It is essential to pay all labourers correctly and on time. The labourers should be informed, well in advance, about the place and date of payment. Once the time and place of payment is agreed, it is very important to stick to your plans. Morale will quickly deteriorate if payment is delayed or incorrect. To make sure payment is correct, you must keep a muster-roll on site where the foreman records labour attendance every day and notes the tasks achieved (see detailed explanation in Chapter 7).

Remember that a lot of preparation is required before pay-day, for example the payment-slips need to be in order and you must arrange with the bank and with security personnel. Some of this can be done well in advance but the payment-slips must be prepared just before payment. To avoid mistakes, and the consequent disputes, sufficient time must be set aside for these preparations. The labourers should also be made aware of the arrangements. Thereby they will understand why they are paid for work done up until the week before payment, but not the last week, so that at the end of the first month they receive payment for only 3 weeks.

# Gravelling Operations

Gravelling consists of a number of operations and activities that follow each other in a step-by-step way:

**Table 8.7** Gravelling operations and activities

| | |
|---|---|
| Improvement of Quarry Access Road (if necessary) | – improvement work to allow the equipment to pass (continuous process during gravelling work) |
| Quarry Preparation | – opening of quarry (removing trees, crops, etc. and removing overburden)<br>– setting out loading bays<br>– excavation and stockpiling of gravel |
| Road Preparation | – reshaping of road according to the desired standards<br>– undertaking small repair work, e.g. washouts, culvert replacement |
| Quarry Work | – excavation and stockpiling of gravel<br>– loading gravel on to hauling equipment |
| Hauling | – hauling gravel from quarry to road |
| Road Reshaping and Gravelling | – reshaping road (if not done before)<br>– off-loading of gravel<br>– spreading and compaction<br>– (provision of gravel stocks for maintenance) |

## IMPROVEMENT OF QUARRY ACCESS ROAD

Where the quarry is not within the road reserve or located directly beside an existing road, an access track will have to be constructed. When you prepare the tender you should carefully assess the work required to build or improve the track so that your lorries or tractors can easily reach the quarry, pass each other, and turn in front of the quarry. Remember that transport equipment is expensive, so it is better to spend money once on a good access road than lose money repeatedly every time a lorry or a tractor is held up.

The gang located at the quarry should make sure that the access road is kept in good condition. A few labourers under the supervision of a gangleader should be delegated for this work. It might also be necessary to provide some loads of gravel at the worst spots once the quarry is established and the equipment is brought to site.

## PREPARATION OF QUARRIES

Normally the client should select and provide the quarry and also make sure that gravel from the chosen quarry meets the specifications. (However, in some contracts it is the responsibility of the contractor to find the quarry and check the quality of the gravel.)

You have to assess carefully the most economic approach if you have a possibility of using more than one quarry. Even if one quarry has sufficient gravel to cover a very long road section, it might be more economical to open another quarry with a shorter haul route. When you assess the costs, remember to include all expenditures involved in opening a new quarry and in improving an additional quarry access road.

Careful planning of the quarry works is needed to have an efficient gravelling operation. The quarry preparation work is carried out well before the gravelling activities start on the road. It is very useful to stockpile sufficient gravel so that once the equipment arrives it can be fully utilized without waiting until enough gravel is excavated.

The quarry layout should allow the tractor and trailers / lorries to enter and leave the quarry without being in each other's way. Possible quarry layouts are shown in Figures 8.6 and 8.7.

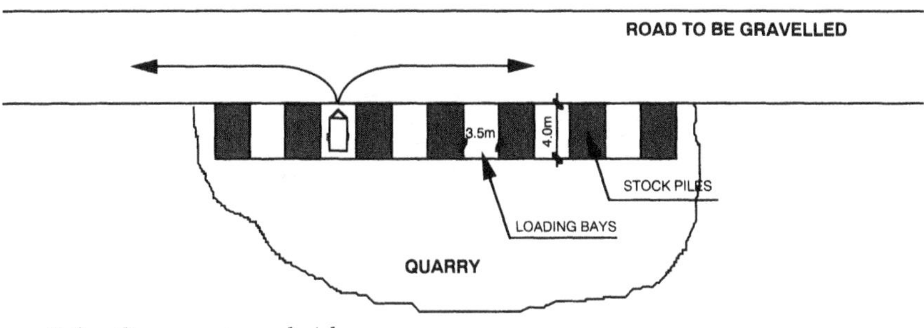


Figure 8.6 *Quarry at road side*

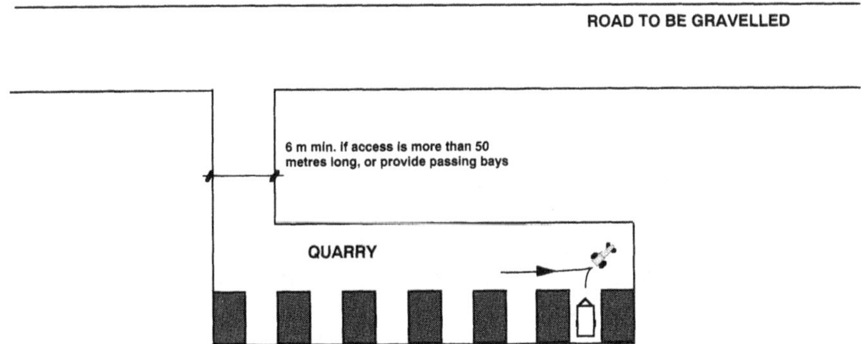


Figure 8.7 *Quarry with access road*

The quarry preparation activities include:

- Setting out the boundaries of the quarry using pegs or boning rods so that the gangleader and labourers know exactly where the quarry is.
- Clearing bush: Before the topsoil can be removed it is necessary to clear the bush and dig out roots. The debris has to be removed from the quarry and, if possible, burnt.
- Excavation of overburden (topsoil): Topsoil must be removed from the area to be quarried, and stockpiled for later use after completing the gravelling operation. Remember always to save the topsoil so that it can be used later to repair the environment and prevent erosion. The activity therefore includes excavation of the topsoil, loading the soil on to wheelbarrows, as well as hauling it to the stockpile area. To haul and place the removed soil back on the quarry site again is of course not a quarry *preparation* activity but it is often useful to include it here when pricing your estimate. The alternative is to make it a separate activity. Either is equally good but remember to include the cost of these replacement activities in your bid.

When excavating the topsoil, make sure that you set out clear task rate areas for all labourers using pegs and strings. To estimate the thickness of the overburden you need to dig some test holes.

**Table 8.8** Proposed task rates for preparation activities

| *Activity* | *Task Rate* | |
|---|---|---|
| Preparation of access road | 50–60m/work-day | |
| Clearing bush | 200–1000m$^2$/work-day | |
| Excavating overburden and loading on to wheelbarrow | 2–4m$^3$/work-day | |
| Hauling overburden by wheelbarrow[1] | Quantity | No. of trips/day |
| 0–40m | 10.5m$^3$/work-day | 210 |
| 40–60m | 8.0m$^3$/work-day | 160 |
| 60-80m | 6.5m$^3$/work-day | 130 |
| 80-100m | 5.5m$^3$/work-day | 110 |

[1] task rate for hauling and tipping only; excludes loading and spreading; two wheelbarrows assigned to each hauling labourer; good haul route.

After the quarry area has been cleared of bushes and overburden, excavation and stockpiling of gravel can start. It is important to excavate and stockpile as much gravel as possible so that the hauling equipment can be fully utilized as soon as it arrives on site. The quarry has to be well established, as shown in Figures 8.6 and 8.7, with clearly defined loading bays for the trailer or lorries to move in (see also guidelines for gravel stockpile in the section on Quarry Work, page 195).

The number of labourers required for the preparation activities depends on the size of the quarry and the volume of work to be carried out over a specified period of time. As a rough guideline it is usually appropriate to assign 40–60 labourers to work in the quarry, including the preparation of the haul route.

Remember: Do *not* bring the hauling equipment before all preparation work is completed and sufficient gravel is stockpiled. From the day the equipment arrives it should be fully utilized in order to contribute as much as possible to your profit. Otherwise you will probably pay a higher rent for the hauling equipment than you can earn from your activities.

When you manage labour and equipment, always remember that the labour force is your 'flexible' resource. You should therefore let your planning be based on equipment availability and utilization. The labour force can then be planned to match the output of the equipment.

## ROAD PREPARATION

In most cases the road to be gravelled has to be brought back to its original cross-section. When light reshaping is sufficient, a length of 20 to 60m per labourer per day is possible. Ditch templates and camberboards must be used to control the reshaping operation. Light reshaping could involve the following activities:

- grubbing of carriageway edge
- reshaping of camber
- minor cleaning of drains and culverts
- shaping of shoulder
- grass cutting in ditch and on shoulder.

You need to be very careful when it comes to estimating the cost of this item. Often the client may specify 'light reshaping' in the tender document but when you visit the road it quite often turns out to be 'heavy reshaping'. If you think that more work is necessary than is specified in the tender documents, you have to notify the client. You may also specify in your bid exactly what you, as a contractor, include under the item 'reshaping of road', if that has not already been done in the client's specifications. However, if you realize the additional work that is necessary only after you have been awarded the contract, it is your duty to notify the client in writing immediately. Explain, in detail, the extra work needed on top of 'light reshaping' and provide a proposal as to how this extra work can be done, and what it would cost. Do not start work until you have a written agreement with the client.

In some cases heavy reshaping or even full rehabilitation is necessary. This book will not deal with this in detail since this type of work is more like construction work than 'normal regravelling'. The contract documents will also need to be much more detailed; almost the same as for construction work.

However, some extra activities over and above light reshaping can always be included in a contract. You can find some guidelines on the construction methods normally employed in Chapter 5. But before you even start planning and pricing any such work, make sure you know exactly what is asked for, and how to carry out such an activity.

Reshaping activities that can be included are:

- cut or re-excavate drains (drainage blocked)
- install new culverts (missing culverts)
- reinstall culverts at correct level or gradient (culvert installed at wrong level or with wrong gradient)
- re-establish drifts/splashes (drift/splash level wrongly set and gradients not sufficient to drain water)
- install scour checks (scour checks are missing on steep gradients and erosion occurs in drains)
- drain widening (in flat area drain has insufficient gradient)
- rebuilding of camber (camber is completely flat and needs to be rebuilt by using material from the side ditch; widening).

These activities are all carried out before the gravelling operation starts. If no compaction is done, sufficient lead-time must be given for traffic and weather to consolidate the carriageway. The necessary length of time, of course, varies with the conditions but a minimum time period would be about two weeks. Where compaction is available reshaping can be done shortly before the gravelling operation.

Always make sure you are well ahead of the gravelling operation, and balance your advance gang to match the speed of the gravelling gang. For example if the gravel team covers 125m a day you must ensure that your reshaping team produces a similar output, or starts earlier in advance.

**Table 8.9** Proposed task rates for road reshaping

| *Activity* | *Task rate* |
|---|---|
| Reshaping road | 20–50m/work-day |

## QUARRY WORK

The quarry preparations are finished when the gravel is cleared of all overburden, an excavation face is opened up and gravel has been stockpiled along the prepared loading bays ready for transportation.

The quarry work can now start and the equipment can move in! Remember what was said about labour being a flexible resource. All labour activities in the quarry and on the road have to be balanced with the expected output of the equipment.

The following quarry activities are to be carried out:

- excavation and stockpiling of gravel
- loading gravel on to trailer or lorry
- hauling the gravel from the quarry to the gravelling site.

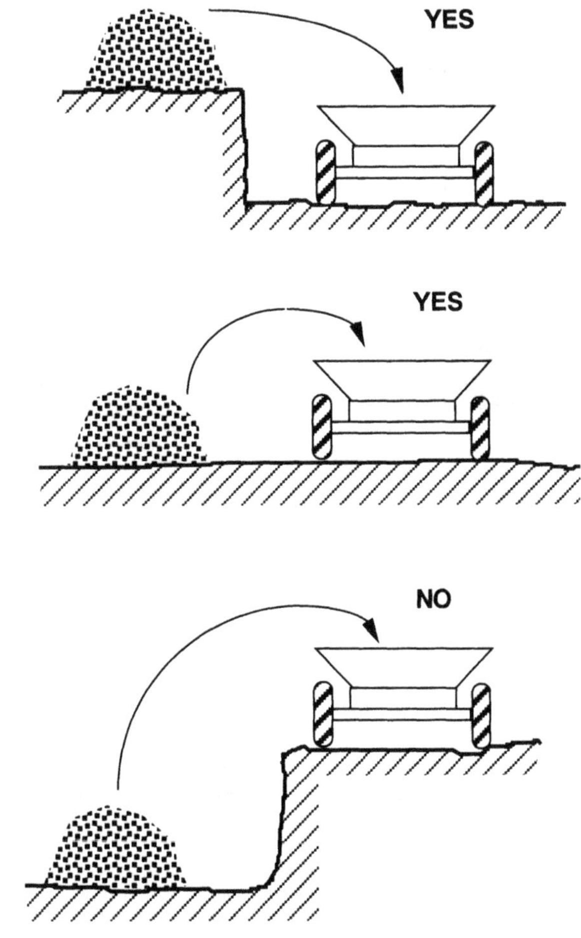


Figure 8.8 *Trailer and lorry loading height*

*Excavation of gravel*

Each day gravel will have to be excavated and stockpiled at the quarry. The quantity stockpiled during a day should match or exceed the amount expected to be hauled the following day. Gravel excavation is suitable for gang tasks. A gang task, compared to individual tasks, reduces the amount of setting-out and monitoring work required by the foremen.

Particular attention should be given to how the quarry develops so that work continues to proceed efficiently. Where possible, bays should be excavated and the gravel stockpiled alongside. In this way the gravel is always cast down into, or from the same level as, the trailers or lorries.

Multiple handling of gravel, moving it from excavation to loading areas and re-stockpiling should be avoided. The following figures (Figures 8.9 and 8.10) show how to organize a quarry ideally.

The excavation to stockpiling activity involves:

- excavating gravel
- breaking any oversize material
- throwing the gravel on to the stockpile.

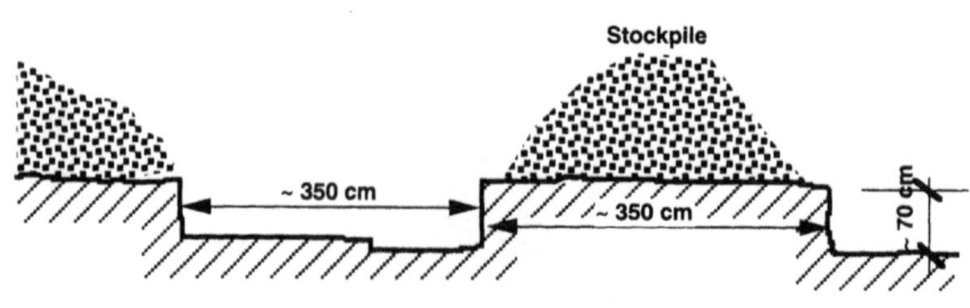


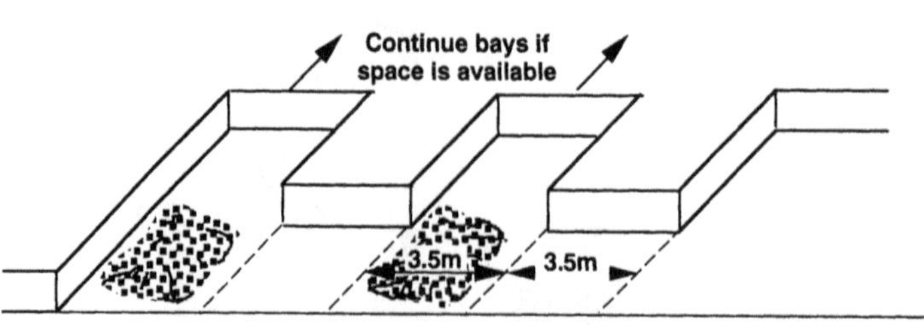


Standard excavation bay about 3.5m wide and 0.7m deep will allow stockpile between bays. Trailer or lorry can be easily reversed into loading bays. A 4m long bay will give about 12m$^3$ of gravel (loose) or 4 trailer-loads of 3m$^3$

Figure 8.9 *Development of quarry on flat land*

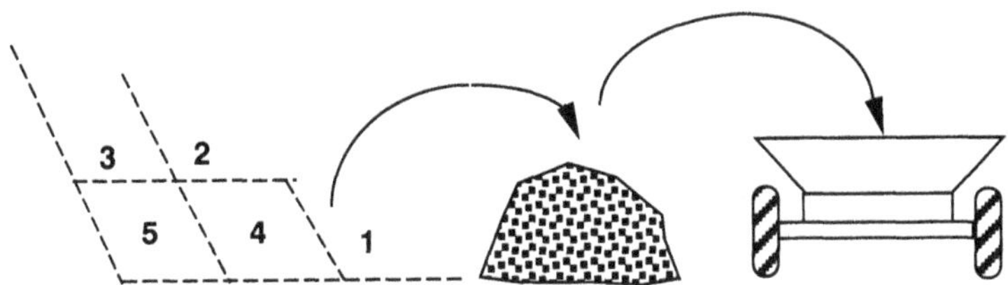


Figure 8.10 *Development of hillside quarry*

Typical labour productivity for this activity is shown in detail by Table 8.10, but you should always make sure your own calculations are based on your own records and experience.

**Table 8.10** Proposed task rates for gravel excavation

| *Activity* | *Task rate* |
|---|---|
| Excavation loose gravel | 2–3m³/work-day |
| Excavation hard gravel | 1.5–2m³/work-day |
| Excavation gravel with oversize | 1–1.2m³/work-day |
| Excavation very hard gravel with much oversize | 1m³/work-day |

*Loading*

The loading gang should be divided into groups of 4–6 labourers. These groups load the empty trailers or lorries as they arrive in the quarry. All trailers or lorries must be fully loaded, level with the body-work.

It is important for you to know exactly the volume that each lorry or trailer can carry. You need this when setting out the off-loading bays at site, and to be able to issue tasks to the loading gang. Average productivity rates for loading are shown in Table 8.11.

**Table 8.11** Proposed task rates for loading gravel

| *Activity* | *Task rate* |
|---|---|
| Loading gravel to tractor trailer | 8–10m³/work-day |
| Loading gravel to lorry | 5–8m³/work-day |

## HAULING

Hauling gravel with tractors and trailers can usually be done using two trailers per tractor to reduce waiting time in the quarry. While one trailer is being loaded in the quarry, the other is pulled by the tractor to the gravelling site. As soon as the tractor returns to the quarry, the empty trailer is parked in a loading bay and a full trailer is hooked on to the tractor. Where there is only one trailer per tractor and tractors have to stand idle in the quarry while the trailer is being loaded, the loading gang should be increased to up to eight labourers per trailer.

Loading lorries also requires 8 to 10 labourers per lorry in order to hold up the lorry as little as possible in the quarry.

The output of the hauling equipment is determined by:

- the availability of the equipment: (how many hours does the equipment work in a day / week / month? This depends mostly on whether you have done proper maintenance on the equipment; if correctly done, there will be few or no breakdowns)
- the speed of the equipment (tractors may travel at an average speed of 15km/h, lorries up to 40 or 50km/h; the condition of the haul route is also important)
- the hauling distance (the longer the distance the fewer trips per day)
- the volume of material that one piece of equipment can carry (a tractor–trailer usually carries $3m^3$ of loose gravel and lorries about 4 to $5m^3$)
- the time required to load and off-load the equipment (your labourers may need up to 20 minutes to load a trailer or lorry and up to 10 minutes to off load it).

*Example:*
*Assume you have access to two tipper trucks, each with a capacity of 4 cubic metres. The hauling distance from the quarry to the road site is about 10km. Therefore it is expected that each tipper truck should do 6 trips per day. The* in situ *quarry material is classified as gravel with some oversize requiring breaking. From the productivity Table 8.10, labour productivity for the excavation is assumed to be $1m^3$/work-day.*

*You can now make the following calculations:*

*a.) The quantity of gravel material that must be excavated each day equals the number of tipper trucks available (A) multiplied by the number of trips each tipper truck will make (B) multiplied by the capacity (C) of each tipper truck, i.e (A x B x C). We get:*

*2 (Trucks) × 6 (Trips) × $4m^3$ (Capacity of truck) = $48m^3$ gravel/day*

*b.) Labour productivity for the excavation has been estimated at $1m^3$/work-day. Therefore the number of labourers required to* *excavate, break oversize and stockpile 48m³ of gravel in one working day is:*

*48m³ divided by 1m³/work-day = 48/1 = 48 labourers*

*c.) The labour productivity for loading tipper trucks has been estimated at 6m³/work-day (see Table 8.11). The number of labourers required to load this material is therefore:*

*48m³ divided by 6m³/work-day = 48/6 = 8 labourers*

To a serious labour-based road contractor this should be a standard calculation that you always undertake to allocate appropriate resources and calculate costs. Equipment costs and labour wages are both major business expenses. By keeping the number of labourers in balance with your transport capacity, you will achieve the best possible productivity. If you use tractors, they should never have to wait in the quarry to be loaded, and the waiting time for lorries must be kept to a minimum.

Standard output tables can be useful, but you must work out your own figures based on your own company's productivity. Make sure that output is checked regularly and that productivity reports are produced daily. With these reports you will see at an early stage if there is a drop in output, and can take action to correct it. In regravelling, with a high cost of equipment, you can lose a lot of money very quickly if productivity goes down. Here are some planning tables that might assist you in developing your own tables, and help you to get started.

For lorries, more than for tractor/trailer combinations, the quality of the haul route is important for the daily productivity. Based on various trials, Table 8.12 has been developed. Use it as a guide to the approximate number of trips one lorry could do per eight-hour working day. But to be able to plan and control *your* work properly you must build up your own rates based on records of past contracts and experience.

**Table 8.12** Hauling planning guidelines (using lorries)

| *Haul route condition* | *Good* | | | | | *Average* | | | | | *Poor* | | | | |
|---|---|---|---|---|---|---|---|---|---|---|---|---|---|---|---|
| Haul distance (km) | 2 | 4 | 6 | 8 | 10 | 2 | 4 | 6 | 8 | 10 | 2 | 4 | 6 | 8 | 10 |
| No. of trips per lorry | 14 | 12 | 11 | 10 | 10 | 13 | 11 | 9 | 8 | 7 | 11 | 9 | 7 | 6 | 5 |

Condition of the haul route:
GOOD: Undulating terrain, good road condition, lorries in good condition
AVERAGE: Hilly terrain, average road condition, trucks in average condition
POOR: Mountainous terrain, poor road condition and difficult quarry access.

**Table 8.13** Gravelling planning guidelines (using tractor–trailer combinations)

| HAUL DISTANCE KM | 45–55hp TRACTOR (EG CASE 533) | | | | | 56–75hp TRACTOR (EG CASE 733) | | | | |
|---|---|---|---|---|---|---|---|---|---|---|
| | TARGET LOADS/ DAY | VOLUME LOOSE $m^3$ | EXCAVATION LABOUR PER TRACTOR | LOADING LABOUR PER TRACTOR | UNLOADING LABOUR PER TRACTOR | TARGET LOADS/ DAY | VOLUME LOOSE $m^3$ | EXCAVATION LABOUR PER TRACTOR | LOADING LABOUR PER TRACTOR | UNLOADING LABOUR PER TRACTOR |
| 0–1 | 18 | 54 | 18–27 | 6–7 | 4–5 | 20 | 60 | 20–30 | 6–8 | 4–5 |
| 1–2 | 14 | 42 | 14–21 | 5–6 | 3–4 | 16 | 48 | 16–24 | 5–6 | 3–4 |
| 2–3 | 10 | 30 | 10–15 | 3–4 | 2–3 | 12 | 36 | 12–18 | 4–5 | 3 |
| 3–4 | 8 | 24 | 8–12 | 3 | 2 | 9 | 27 | 9–14 | 3–4 | 2–3 |
| 4–5 | 6 | 18 | 6–9 | 2–3 | 2 | 8 | 24 | 8–12 | 3 | 2 |
| 5–6 | 5 | 15 | 5–8 | 2 | 1–2 | 6 | 18 | 6–9 | 2–3 | 2 |
| 6–7 | 5 | 15 | 5–8 | 2 | 1–2 | 5 | 15 | 5–8 | 2 | 1–2 |
| 7–8 | 4 | 12 | 4–6 | 2 | 1 | 5 | 15 | 5–8 | 2 | 1–2 |
| 8–9 | 3 | 9 | 3–5 | 1–2 | 1 | 4 | 12 | 4–6 | 2 | 1 |
| 9–10 | 3 | 9 | 3–5 | 1–2 | 1 | 3 | 9 | 3–5 | 1–2 | 1 |

ACCEPTABLE RANGE: TARGETS PER TRACTOR

| HAUL DISTANCE KM | 45–55hp TRACTOR (EG CASE 533) | | 56–75hp TRACTOR (EG CASE 733) | |
|---|---|---|---|---|
| | TARGET LOADS PER DAY | | TARGET LOADS PER DAY | |
| | POOR HAUL ROUTE | GOOD HAUL ROUTE | POOR HAUL ROUTE | GOOD HAUL ROUTE |
| 0–1 | 16 | 19 | 18 | 21 |
| 1–2 | 12 | 16 | 14 | 18 |
| 2–3 | 8 | 11 | 10 | 13 |
| 3–4 | 6 | 9 | 8 | 10 |
| 4–5 | 5 | 7 | 6 | 9 |
| 5–6 | 4 | 6 | 5 | 7 |
| 6–7 | 4 | 5 | 4 | 6 |
| 7–8 | 3 | 5 | 4 | 6 |
| 8–9 | 3 | 4 | 3 | 5 |
| 9–10 | 2 | –4 | 3 | 4 |

ASSUMPTIONS:
- 3 CUBIC METRE TRAILERS, 8 HOUR DAY, 1 HOUR TRACTOR SERVICE TIME/DAY, 7 HOURS TRACTOR WORKING TIME/DAY, 3 MINUTES TO CHANGE TRAILERS AT QUARRY, 8 MINUTES TO OFF LOAD TRAILER.
- POOR HAUL ROUTE: ALTITUDE > 2500 METRES, GRADIENTS > 10%, POOR HAUL SURFACE.
- TRACTOR/TRAILER HAULS OF MORE THAN 8 KMS ARE UNECONOMIC AND SHOULD BE AVOIDED.
- HAUL LENGTHS OF MORE THAN 10 KM SHOULD BE DISCUSSED WITH THE REGIONAL COORDINATOR.
- FOR PRACTICAL REASONS THE UNLOADING GANG SHOULD NORMALLY CONSIST OF 4 OR MORE PERSONS.
- LABOUR SHOULD NOT BE ASSIGNED TO A PARTICULAR TRACTOR: THE LABOUR REQUIREMENTS SHOULD BE CALCULATED FOR THE TOTAL NUMBER OF TRACTORS IN USE AND THE DAILY QUANTITIES BEING HAULED.

## GRAVELLING

The actual gravelling operation consists of four main activities:

- ○ reshaping the road (already discussed in the section on Road Preparation)
- ○ off-loading the gravel
- ○ spreading the gravel
- ○ compacting the fresh gravel layer.

When gravelling a road you can either work:

1.) towards the quarry, or:

Figure 8.11a

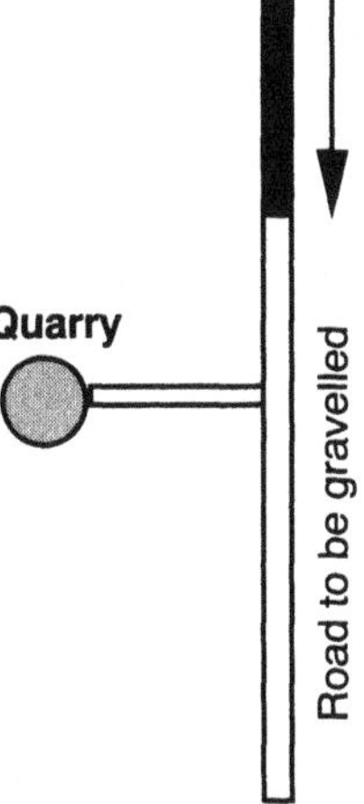


2.) away from the quarry:

Figure 8.11b

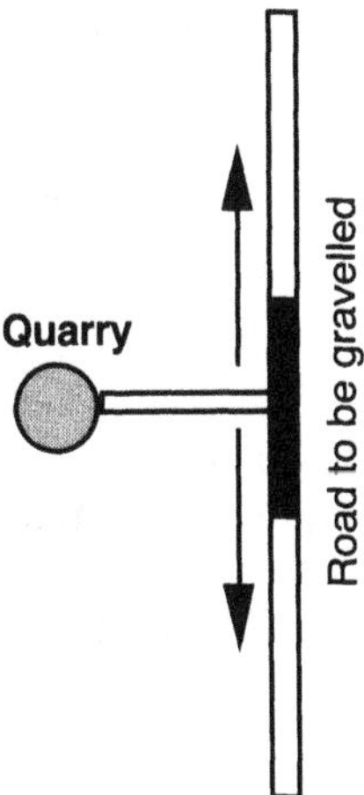

We recommend that you commence gravelling from where the quarry access road joins the road to be gravelled (example No 2). Initially the road should be gravelled away from the quarry access *in both directions* at the same time. As you have short hauls you can reduce congestion at the unloading sites. When the hauls exceed approximately 1km, gravelling should continue in one direction only. You need to continue with one direction at a time to ease supervision and utilize unloading/spreading gangs as much as possible. If you follow this proposed method you can achieve the following advantages:

- The vehicles compact the material as they haul over the gravel already laid.
- Damage to earth road camber is minimized.
- Gravelling traffic does not interfere with reshaping activities.
- Gravelling can recommence quicker after rainfall.

*Off-Loading and Spreading of Gravel*

If the hauling equipment has no tipping mechanism, off-loading is done by labour. Off-loading and spreading are usually combined in one activity, carried out by one gang of labourers. It is important that the trailers/lorries are off-loaded as quickly as possible. However, no more than four to six labourers can work on a trailer or a lorry at the same time without hindering each other.

The tractor – trailer / lorry should turn, if possible, before unloading. In this way it can return to the quarry without waiting for the material to be spread.

The material from one trailer/lorry is unloaded and spread within a 'box' marked by pegs and strings. The width of the box is the road's running-surface width (specified by standard cross-section in contract) while you have to calculate the length. You base your calculations on the thickness of loose gravel material that is needed to fulfil the contract specifications.

$$\text{The box length} = \frac{\text{Trailer/lorry capacity}}{\text{Road width} \times \text{layer thickness (loose)}}$$

The gravel thickness in the contract is specified as 'compacted'. You need to calculate the required thickness of loose gravel to get the specified compacted layer. Usually you have to add a factor of 25% (add 25% = multiply by 1.25).

For example, the gravel thickness is specified as 150mm (15cm) in the contract; to calculate the required thickness of loose gravel you multiply by 1.25.

150mm x 1.25 = 190mm (exact result 187.5mm)

However, the factor of 25% varies, depending on several factors such as the quality of gravel and the moisture content. It is therefore necessary to measure closely the actual achieved gravel thickness after compaction. You may have to allow for some minor corrections.

Pegs should first be set at the centre line (establish the centre line well in advance of the gravelling work). Then you can measure the width of the road from there and set the side pegs. Strings should be fixed at the height that the material should reach when evenly spread. This helps to ensure that the gravel layer has a uniform thickness.

*Example:*
*The contract specifies a surface width of 400cm with a compacted gravel layer of 15cm. In order to get the required thickness for loose material we add 25% to 15cm, which gives us 19cm.*

*Now we need to define the length of the 'box'. It is the length which can be filled by one load of our 4m³ lorries. The calculation is as follows:*

$$\text{'box' length, } L = \frac{\text{Lorry (or trailer) capacity}}{\text{Road width} \times \text{layer thickness (loose)}}$$

*Lorry capacity* = $4m^3$
*Road width* = *4m*
*Layer thickness (loose)= 19cm = 0.19m*

$$L = 4 / [4 \times 0.19] = 5.26m$$

*To set out boxes with a length of 5.26m length would probably be rather impractical in the field. We would therefore tell our gangleader to set out boxes with a length of 5.30m and tell the gangleaders of the loading teams to fill the trailers a little bit above the normal level.*

*Our boxes look as shown in Figure 8.12.*

To avoid delaying the next load of material, the spreading gang has immediately to spread the dumped material. Watering and breaking of oversize material can take place after the load is spread. Always check with the camberboard or a straight edge and spirit-level that the right camber gradient is achieved.

This activity is usually best issued as a gang task. To organize work efficiently you should place a dynamic and flexible person as your gangleader. He or she must be capable of controlling arriving vehicles as well as labourers in what sometimes are very hectic situations.

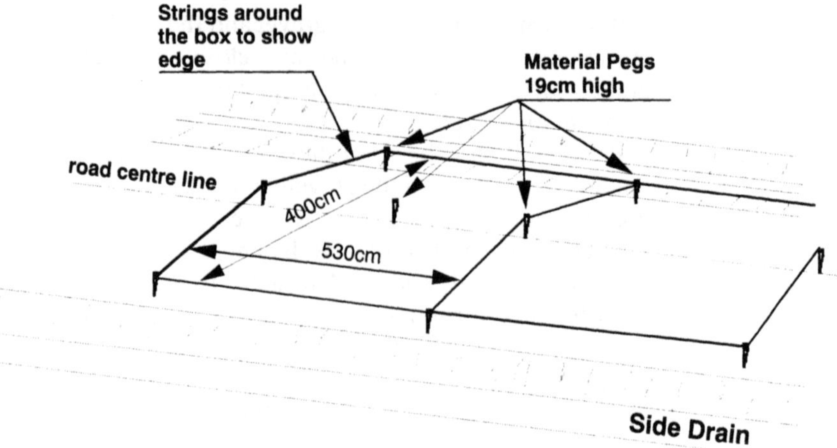


Figure 8.12 *Gravel boxes*

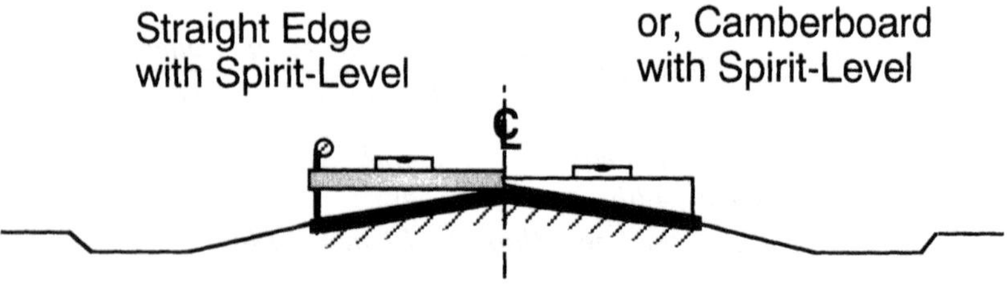


Figure 8.13 *Checking of camber gradient*

**Table 8.14** Proposed task rates for off-loading and spreading

| *Activity* | *Task rate* |
|---|---|
| Off-loading and spreading | 12–16m³/work-day |

Finally, do not leave a step along the gravel layer edges. During spreading and subsequent compaction, allow some gravel material to spill over on to the shoulders but only as far as the edges of the roadway. Do not waste gravel; any spillage into the drains must be thrown back on to the gravelled area.

## COMPACTION

As gravelling starts at the quarry, the hauling equipment will drive over the fresh gravel layer. This gives a certain initial compaction. To ensure an even compaction of the running surface, the drivers should be instructed not always to follow the same track. They should start to drive on one side of the road and move across towards the centre of the road with each subsequent pass of the loaded trailer or lorry. This procedure should then be repeated for the other side of the road. During this process small ruts are usually formed. To ensure a uniform surface you should leave one or two labourers behind on the completed sections to carry out continuous reshaping of the gravel layer.

Where the contract specifies compaction, you must use compaction equipment. Dead weight rollers towed by a tractor can often achieve a satisfactorily result. Alternatively you might need pedestrian vibrating rollers. They have become standard equipment on labour-based road projects, where they have achieved good results. Usually some six to eight passes with the roller is sufficient to achieve the required compaction.

Water needs to be added where the natural moisture content of the gravel is below the optimum. The best equipment for this is often a tractor-towed water bowser with a spraying bar.

The compaction activity must be well organized so that you can always compact the gravel spread on the same day. Where you need to add water your plans must include transport of water from the source to your site. This may reduce your output on gravel hauling, as one of your tractors or lorries will be tied up for a period of time.

# Planning and Monitoring Gravelling Work

## ADVANCE PLANNING OF WORK

The advance planning for gravelling is usually done by the client, and includes:

- estimating the amount of gravel needed, based on the carriageway width, average gravel thickness, and length of gravelling
- the identification and assessment of quarries, including haul distance for each section of road.

The contractor will get this data from the tender document. Before you can fill out the tender you have to visit the road for a personal assessment. A sketch of the road, as shown in our example in Figure 8.2, can be very useful for this purpose.

Based on the information provided in the tender document, your site inspection, and your experience you will then be able to:

- estimate the number of work-days required for quarry preparation, reshaping and gravelling
- estimate the number of equipment-days required to haul the gravel from the quarry to site.

*Example:*
*The contract specifies a surface width of 400cm with a compacted gravel layer of 15cm. In order to get the required thickness of loose material we add 25% to 15cm, which gives us 19cm (rounded up). Per metre road length we therefore require 0.76m*$^3$ *of loose gravel (width × thickness × length = 4.0m × 0.19m × 1.0m = 0.76m*$^3$*)*

*The requirements for each section (see Figure 8.2) are:*

*Quarry I: From our site visit notes we know that Quarry I is 30 × 50m in size and that the overburden is 30cm as an average.*

- *length to be gravelled = 5500m (Figure 8.2)*
- *loose gravel required = length × volume of gravel per metre of road 5500 × 0.76 = 4180m*$^3$
- *work-days required:*

*The work-days required are calculated using the tables of task rates previously presented in this book (Tables 8.8 – 8.11)*

*Quarry preparation; length of access road = 2150m (Figure 8.2)*

- *access road improvement (Table 8.8) =*
  *2150m/60m per workday = 36wd*
- *bush clearing (Table 8.8) = 1500m²/500m² p.wd = 3wd*
- *overburden removal (Table 8.8) = 450m³/3m³ p.wd = 150wd*

*Road reshaping:*

- *reshaping (Table 8.9) = 5500m/40m p.wd = 138wd*

*Gravelling:*

- *excavation of hard gravel (Table 8.10) =*
  *4180m³/1.7m³ p.wd = 2459wd*
- *loading to lorry (Table 8.11) = 4180m³/7m³ p.wd = 597wd*
- *off-loading + spreading (Table 8.12) =*
  *4180m³/15m³ p.wd = 279wd*
- *miscellaneous (support, reshaping, etc.) + 10% = 366wd*

***Total Quarry I*** ***4028wd***

*Quarry II:*
- *length to be gravelled = 6500m (Figure 8.2)*
- *loose gravel required = 4940m³ (6500 x 0.76)*
- *work-days required:*

*Quarry preparation; length of access road = 4350m (Figure 8.2)*

- *access road improvement (Table 8.8) =*
  *4350m/60m p.wd = 73wd*
- *bush clearing (Table 8.8) = 3500m²/300m² p.wd = 12wd*
- *overburden removal (Table 8.8) =*
  *1050m³/3m³ p.wd = 350wd*

*Road reshaping:*

- *reshaping (Table 8.9) = 4940m/40m p.wd = 124wd*

*Gravelling:*

- *excavation with oversize gravel (Table 8.10) =*
  *4940m³/1.2m³ p.wd = 4117wd*
- *loading to lorry (Table 8.11) = 4940m³/7m³ p.wd = 706wd*
- *off-loading + spreading (Table 8.12) =*
  *4940m³/15m³ p.wd = 329wd*
- *miscellaneous (support, reshaping, etc.) + 10% = 571wd*

***Total Quarry II*** ***6282wd***

*Quarry III:*
- *length to be gravelled = 3000m (Figure 8.2)*
- *loose gravel required = 2280m³ (3000 x 0.76)*
- *work-days required:*

*Quarry preparation; length of access road = 630m*

- *access road improvement (Table 8.8) =*
  *630m/60m p.wd = 11wd*
- *bush clearing (Table 8.8) = 800m²/300m² p.wd = 3wd*
- *overburden removal (Table 8.8) = 320m³/3m³ p.wd = 107wd*

*Road reshaping:*

- *reshaping (Table 8.9) = 3000m/40m p.wd = 75wd*

*Gravelling:*

- *excavation with oversize gravel (Table 8.10) =*
  *2280m³/1.2m³ p.wd = 1900wd*
- *loading to lorry (Table 8.11) = 2280m³/7m³ p.wd = 326wd*
- *off-loading + spreading (Table 8.12) =*
  *2280m³/15m³ p.wd = 152wd*
- *miscellaneous (support, reshaping, etc.) + 10% = 258wd*

***Total Quarry III*** ***2832wd***

***Total workerdays required all quarries*** *(4028 + 6282 + 2832)*
***13 142wd***

*The equipment-days (in our case lorries) will be based on the above quantity of gravel to be transported and the average hauling distance for each section of road:*

*Quarry I:*
*Average hauling distance = 3.525km*
*Amount of gravel to be hauled = 4180m³*
*Lorry capacity = 4m³*

*Total lorry loads required 4180/4 = 1045 loads (trips)*

*Trips per lorry per day (Table 8.13, good haul route) = 12 trips*

***Total required lorry-days for Quarry I (1045/12) =***
***87 lorry-days***

*Quarry II:*
*Average hauling distance = 7.090km*
*Amount of gravel to be hauled = 4940m³*
*Lorry capacity =4m³*
*Total lorry loads required (4940/4) = 1235 loads (trips)*
*Trips per lorry per day (Table 8.12, average haul route) = 8 trips*

***Total required lorry days for Quarry II (1235/8) =***
***155 lorry-days***

*Quarry III:*
*Average hauling distance = 1380km*
*Amount of gravel to be hauled = 2280m³*
*Lorry capacity = 4m³*

*Total lorry loads required (2280/4) = 570 loads (trips)*

*Trips per lorry per day (Table 8.12, good haul route) = 14 trips*

***Total required lorry-days for Quarry III (570/14) = 41 lorry-days***

***Total lorry-days required all quarries (87 + 155 + 41) 283 lorry days***

This now gives you a first indication of the required manpower and equipment. Based on your decision on how many pieces of equipment you will be using for a particular contract you can then estimate the time required for the job.

*Example:*
*In our example we estimated a total of 283 lorry-days. The actual time required to haul the gravel to site now depends on the number of lorries we will assign for the job. In our case we had initially thought of using only our own two lorries. We would therefore require about 142 days (or 6.5 months) just to transport the gravel.*

*However, the contract specifies that the total time to be spent on the contract should not exceed 4 months. We also have to allow some time to carry out all the preparatory works and to excavate and stockpile sufficient gravel. For these activities we plan one month. We are therefore left with 3 months for the actual gravelling job. The average working days per month we are estimating to be 22 which gives us 66 working days. The lorry-days required are 283, and divided by 66 working days we get 4.3 lorries. This means that we require 5 lorries to finish the job in time. Using 5 lorries will also be realistic when we consider that there will be some break-downs or other interruptions which will not always allow us to achieve the daily target.*

*In our case we have decided to hire three additional lorries for three months, from another company, to complement our fleet of two.*

## SITE OPERATION PLANNING

Once you have been awarded a contract, you have to prepare the detailed site operation plans. This is best done on a weekly basis, as the requirements will keep changing from week to week.

First comes the preparation activities done before you can move in with your hauling equipment. On the basis of calculations similar to those in the example above, you should know the approximate number of work-days required. You also have to add the work-days that are required for the excavation and stockpiling of gravel until the equipment arrives. This depends very much on the size and layout of the quarry.

*Example:*
*Quarry I is 30 × 50m in size. For the total preparatory activities we had estimated 189 work-days plus 10% for support, etc. which gives us a total of approximately 210 work-days.*

*We have also estimated the maximum excavation and stockpiling possible in this quarry before the hauling starts as follows:*

- *front length of quarry = 50m*
- *width of bays for lorries = 3.5m*
- *the quarry has a slope to the back of only about 3% so the bays can be dug into the hill with a uniform height of 0.7m. It is therefore possible to dig the bays from the front of the quarry right through to the back.*
- *in total it will be possible to have 14 bays (50m / 3.5m) of which seven can be excavated while the remaining seven will be used to stockpile the gravel.*
- *the total volume of gravel that can be excavated per bay is 3.5m × 30m × 0.7m = 73.5m$^3$ compacted, or about 92m$^3$ loose.*
- *the total volume of loose gravel that can be excavated and stockpiled prior to the arrival of the lorries is therefore 92m$^3$ x 14 bays = 1288m$^3$*
- *we estimated the productivity for this quarry to be 1.7m$^3$ / work-day and we therefore require 758 work-days to excavate and stockpile the gravel, plus some 10% for support, etc. which gives us a total of approximately 840 work-days.*

*Including the work-days for preparatory activities we therefore have to plan for a total of 1050 work-days for the one month that is allowed to us for the preparatory activities. Assuming that this particular month has 23 working days we will require an average of 46 labourers per day (1050 wd / 23days = 46 labourers). As there will always be some deserters or labourers that cannot finish their task we will initially employ a labour force of 55 labourers. That also gives us the opportunity to plan work in a staggered way as it will not be possible to start work with the full labour force on the first day. The camp has to be established, the access road repaired and the bush cleared on the quarry with a gang of 15 labourers for one week before the remaining 40 labourers can be called in for overburden removal and excavation. This gives us a potential total of 1065 work-days (15 labourers × 5 days = 75 wd, plus 55 labourers × 18 days = 990 wd; 75 + 990 = 1065).*

After this preparatory phase you can now start to bring in your equipment and haul the gravel. The planning from now on is based on the capacity of the equipment. Whatever amount of gravel the equipment can transport has to be stockpiled ready for loading, and at the other end needs to be off-loaded, spread and compacted.

For this purpose daily planning is required. The form shown in Figure 8.14 may assist you in the planning process. The same form can also be used to record the actual achieved output to compare it with the set target. You may have to adjust the form to suit your particular needs. It is important that you instruct your site supervisor how to fill out the form. It is, however, advisable that you carry out the planning yourself while the site supervisor fills in the reports. The planning has to be done very carefully so that you do not waste any resources.

The basic procedure for preparing the plan is:

- determine the haulage capacity available
- determine the haulage distance and gravel haulage target
- verify the availability of gravel stockpiles
- allocate labour to each activity according to requirements
- cross-check the total labour requirements assuming usual attendance
- adjust the number of items of gravelling equipment or labour gangs as necessary.

As you will have more than one gang it will be necessary to issue each gang with a separate form. For control and monitoring purposes you can then transfer the data on to one form and to a bar-chart. A bar-chart will assist you to get an overview of the different activities and the progress achieved compared to the targets previously set.

The filled-out form (Figure 8.15) is based on our example and reflects one week during the time when the project is in full operation.

The reshaping of the road should start well before the hauling operations. If there are just minor reshaping activities needed you can begin this work about two weeks before the gravelling. For major reshaping (or even rehabilitation), you will probably need to start earlier. The gang size depends on the speed you want to achieve, and the work-load required.

*Example:*
*In our case, the road is still in good shape and relatively little work needs to be done prior to the gravelling activities. The camber has to be repaired in some places, the ditch has some silt pockets and a few culverts need to be cleaned. We therefore assume that the daily productivity could be on average 50m per labourer.*

*We have assumed the output to come from an average of 4.5 lorries (we have 5), allowing for minor break-downs etc. The average daily gravelling output from Quarry I will be some 284m (12 trips for each of the 4.5 lorries and each trip producing 5.26m surface layer gives a total of 284m).*

*With this daily reshaping productivity we need a labour gang of 6 labourers (284m/50m = 5.7 per day) to match the speed of the gravelling team. As we are gravelling away from the quarry, and we do not want the lorries to get crowded, we start gravelling to both sites. This means that we will deploy three labourers for reshaping to each side. They will commence work 2 weeks before the gravelling team starts so that the road can consolidate in the meantime.*

As soon as the equipment can start to haul gravel you have to make sure your labour gangs are well balanced, enabling them to cope with the pace of the equipment. Remember, the loading gang, the spreading gang and the excavation gang must be balanced so that your equipment never stands idle.

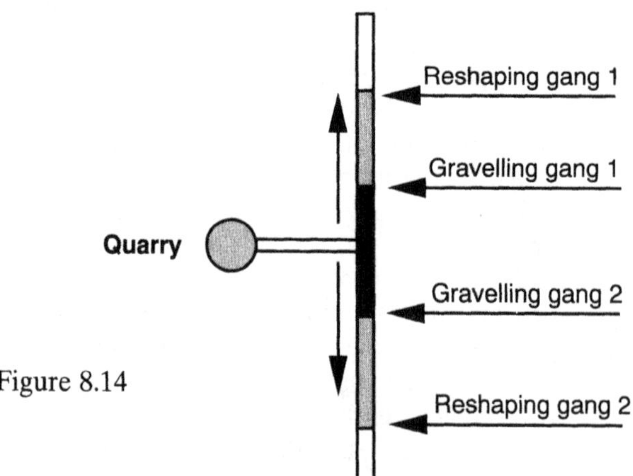


Figure 8.14

*Example:*
*We assume an average of 4.5 lorries, each undertaking 12 trips a day from quarry I to the two gravelling sites. In order to have both sites balanced, it will be necessary to assign two lorries to each site and one additional lorry to alternate between the two*

## WEEKLY GRAVELLING PLAN/ REPORT

| | | |
|---|---|---|
| ROAD: | QUARRY: | PERIOD FROM: TO: |
| HAUL DISTANCE THIS WEEK: KM | NO. OF TR. / L. | NO OF TRIPS / TR. OR L. (A) |
| TRAILER/LORRY CAPACITY (B) M3 | GRAVEL WIDTH (C) M | THICKNESS LOOSE (D) M |
| SPREAD LENGTH (E) = B/(CXD) M | DATE: | PREPARED BY: |

| EQUIPMENT USE | DATE | | | | | | | | |
|---|---|---|---|---|---|---|---|---|---|
| | DAY | MON | TUE | WED | THU | FRI | SAT | SUN | Total |
| NUMBER OF TR / L ON SITE (F) | NO | | | | | | | | |
| NUMBER OF TRIPS (G) = A X F | NO | | | | | | | | |
| SECTION LENGTH GRAVELLED = G x E | M | | | | | | | | |

| | ACTIVITIES | TASK RATE | DATE | | | | | | | | |
|---|---|---|---|---|---|---|---|---|---|---|---|
| | | | DAY | MON | TUE | WED | THU | FRI | SAT | SUN | Total |
| QUARRY PREPARATION | QUARRY ACCESS ROAD | M | Quantity | | | | | | | | |
| | | | Gang Size | | | | | | | | |
| | BUSH CLEARING | M2 | Quantity | | | | | | | | |
| | | | Gang Size | | | | | | | | |
| | OVERBURDEN REMOVAL | M3 | Quantity | | | | | | | | |
| | | | Gang Size | | | | | | | | |
| | OVERBURDEN HAULING | M3 | Quantity | | | | | | | | |
| | | | Gang Size | | | | | | | | |
| | GRAVEL EXCAVATION | M3 | Quantity | | | | | | | | |
| | | | Gang Size | | | | | | | | |
| | QUARRY BACKFILLING | M3 | Quantity | | | | | | | | |
| | | | Gang Size | | | | | | | | |
| | SUPPORT | | | | | | | | | | |
| | | | Gang Size | | | | | | | | |
| GRAVELLING | ROAD RESHAPING | M | Quantity | | | | | | | | |
| | | | Gang Size | | | | | | | | |
| | GRAVEL EXCAVATION | M3 | Quantity | | | | | | | | |
| | | | Gang Size | | | | | | | | |
| | LOADING | M3 | Quantity | | | | | | | | |
| | | | Gang Size | | | | | | | | |
| | OFFLOADING + SPREADING | M3 | Quantity | | | | | | | | |
| | | | Gang Size | | | | | | | | |
| | OTHERS | | Quantity | | | | | | | | |
| | | | Gang Size | | | | | | | | |
| | SUPPORT | | | | | | | | | | |
| | | | Gang Size | | | | | | | | |
| | TOTAL WORK-DAYS | | | | | | | | | | |

TR. = TRACTOR L. = LORRY

Figure 8.15 *Weekly gravelling plan/report*

## WEEKLY GRAVELLING PLAN/ REPORT

| ROAD: *EXAMPLE* | QUARRY: *QUARRY I* | PERIOD FROM: *4.7.* TO: *8.7.95* |
|---|---|---|
| HAUL DISTANCE THIS WEEK: *3.525* KM | NO. OF TR. / L. *4.5* | NO OF TRIPS / TR. OR L. (A) *12* |
| TRAILER/LORRY CAPACITY (B) *4.0* M3 | GRAVEL WIDTH (C) *4.0* M | THICKNESS LOOSE (D) *0.19* M |
| SPREAD LENGTH (E) = B/(CXD) *5.26*M | DATE: *3.7.95* | PREPARED BY: *X.Y.* |

| EQUIPMENT USE | DATE | *4.7* | *5.7* | *6.7* | *7.7* | *8.7* | | | |
|---|---|---|---|---|---|---|---|---|---|
| | DAY | MON | TUE | WED | THU | FRI | SAT | SUN | Total |
| NUMBER OF TR / L ON SITE (F) | NO | *4.5* | *4.5* | *4.5* | *4.5* | *4.5* | | | |
| NUMBER OF TRIPS (G) = A X F | NO | *54* | *54* | *54* | *54* | *54* | | | *270* |
| SECTION LENGTH GRAVELLED = G x E | M | *284* | *284* | *284* | *284* | *284* | | | *1420* |

| | ACTIVITIES | TASK RATE | DATE | | | | | | | | |
|---|---|---|---|---|---|---|---|---|---|---|---|
| | | | DAY | MON | TUE | WED | THU | FRI | SAT | SUN | Total |
| QUARRY PREPARATION | QUARRY ACCESS ROAD | M | Quantity | | | | | | | | |
| | | | Gang Size | | | | | | | | |
| | BUSH CLEARING | M2 | Quantity | | | | | | | | |
| | | | Gang Size | | | | | | | | |
| | OVERBURDEN REMOVAL | M3 | Quantity | | | | | | | | |
| | | | Gang Size | | | | | | | | |
| | OVERBURDEN HAULING | M3 | Quantity | | | | | | | | |
| | | | Gang Size | | | | | | | | |
| | GRAVEL EXCAVATION | M3 | Quantity | | | | | | | | |
| | | | Gang Size | | | | | | | | |
| | QUARRY BACKFILLING | M3 | Quantity | | | | | | | | |
| | | | Gang Size | | | | | | | | |
| | SUPPORT | | | | | | | | | | |
| | | | Gang Size | | | | | | | | |
| GRAVELLING | ROAD RESHAPING | M | Quantity | *300* | *300* | *300* | *300* | *300* | | | *1500* |
| | | *50* | Gang Size | *6* | *6* | *6* | *6* | *6* | | | *30* |
| | GRAVEL EXCAVATION | M3 | Quantity | *127.5* | *127.5* | *127.5* | *127.5* | *127.5* | | | *637.5* |
| | | *1.7* | Gang Size | *75* | *75* | *75* | *75* | *75* | | | *375* |
| | LOADING | M3 | Quantity | *216* | *216* | *216* | *216* | *216* | | | *1080* |
| | | *6.96* | Gang Size | *31* | *31* | *31* | *31* | *31* | | | *155* |
| | OFFLOADING + SPREADING | M3 | Quantity | *216* | *216* | *216* | *216* | *216* | | | *1080* |
| | | *15.4* | Gang Size | *14* | *14* | *14* | *14* | *14* | | | *70* |
| | OTHERS | | Quantity | | | | | | | | |
| | | | Gang Size | *2* | *2* | *2* | *2* | *2* | | | *10* |
| | SUPPORT | | | | | | | | | | |
| | | | Gang Size | *8* | *8* | *8* | *8* | *8* | | | *40* |
| TOTAL WORK-DAYS | | | | *136* | *136* | *136* | *136* | *136* | | | *680* |

TR. = TRACTOR L. = LORRY

Figure 8.16 *Weekly gravelling plan/report (filled out example)*

*sites. This way both sites will receive some 27 loads of gravel, which allows them to gravel about 142m.*

*In order to be able to off-load and spread the gravel on each site (each site receives 27 loads of $4m^3 = 27 \times 4 = 108m^3$) we require 7 labourers ($108m^3 / 15m^3$ per work-day). For both sites we therefore require 14 labourers.*

*The total labour requirement for reshaping and off-loading is 20 labourers.*

*For loading the 54 lorry-loads we require another gang of 30 labourers (54 lorries x $4m^3 = 216m^3$; $216m^3 / 7m^3$ per work-day = 31 labourers).*

*In order to make sure we can load the lorries as quickly as possible, we form three gangs of 10 labourers, each gang working in a separate loading bay.*

*For the excavation we require another labour gang of approximately 70 to 80 labourers.*

*Their output will not be at a level to match the daily equipment output but we have already stockpiled some $1288m^3$ (or 322 lorry-loads) of gravel prior to the arrival of the lorries. We also foresee some breakdown or days where not all trips can be done because of bad weather. On those days we will allocate surplus labour from the activities to excavation.*

***The total labour force will therefore be between 120 and 130 labourers. (20 + 30 + 70 = 120)***

From the example it is easy to see that it will be necessary to replan almost every day as the requirements will keep changing. It is therefore very important that you carefully assess every day's performance and base your planning for the next day on the actual requirements and the equipment capacity available.

The planning process explained in this example covers only one quarry. The same steps need to be carried out for each of the other quarries. Remember, you need to start quarry preparation and excavation of gravel in Quarry II early enough to utilize the equipment fully when it moves from the first to the second quarry. This probably means that you will have to employ more labourers temporarily. On the other hand, when the work is coming to an end in the first quarry you will start to lay off labourers there. After a period of time you will again have a balanced labour-force. When handling these fairly large groups of labour it is easy to understand the importance of estimating your labour requirements well in advance. An additional practical aspect is the necessity to make arrangements for labour payments.

# CHAPTER 9: PRICING AND BIDDING

## Learning Objectives

After you have completed this chapter you should know:

- how to deal with tenders and contracts
- how to develop and estimate the prices for routine maintenance items
- how to develop and estimate the prices for regravelling items.

## Introduction

Many small-scale contractors have serious problems surviving. Experience from several countries tells us that one of the most important factors contributing to these problems is the contractor's inability to assess costs and prepare a proper bid. A typical small-scale contractor, with limited financial resources, needs to have new projects coming in at fairly regular intervals and must make a reasonable profit on every project undertaken. Putting in bids that are too high means not getting any projects, while just one bid that is far too low might lead to serious financial problems.

To be able to prepare a bid, you need information on the project (bidding documents plus site inspection), and your own company (company records, follow-up of previous contracts). However, the most important input is always your own experience (for example: what profit margin to add, what are the risks involved, when will I get paid?).

Being such an important activity, it is obvious that preparation of bids cannot normally be delegated to any of your staff. Only you know all the required details and what assumptions have to be made, so only you can assess the risk a project poses to your company and estimate the level of profit to be added. However, you can only undertake this properly if you have complete and up-to-date information on your company and its activities. Productivity rates, based on previous projects, for all activities, a list of current prices for materials, and properly-calculated equipment costs are all examples of input needed.

This chapter gives you guidelines on how to calculate prices and prepare bids for routine maintenance and regravelling contracts. The principles applied follow the system introduced in IYCB Handbook 1, 'Pricing and Bidding' and the charts developed for the IYCB books are used extensively in this chapter. It also discusses the importance of proper cost calculations and how they provide you with tools to manage the finances of your firm, for example, 'cash-flow forecasts'. A general background on how to estimate costs and calculate prices is provided in IYCB Handbook 1, 'Pricing and Bidding'. We suggest that you look through the IYCB book before you read this chapter, and for more information on cash-flow and how to prepare it, see IYCB Handbook 3, 'Business Management'.

## Contract Management

Mastering tender and contract procedures is not an easy task; it covers such a wide range of knowledge. To take all relevant items and aspects properly into consideration is complicated, and it is easy to forget or overlook some aspects.

In order to facilitate your work we have prepared a checklist. This checklist covers the most important items you might come across when preparing and negotiating bids. Please note that some of these items might not be applicable to the particular tender or contract you are dealing with.

- Contractor registration with the client
- Responding to tender invitation
- Study the tender documents
- Site visit
- Cost estimating
  - Direct costs
  - Indirect costs
  - Profit
  - Contingencies
- Financing
- Tender submission
- Contract negotiations
- Mobilization
- Work management
- Payment certificates
- Monitoring performance
- Training and manpower development
- Disputes and arbitration

## CONTRACTOR REGISTRATION WITH CLIENT

In most cases the contractor needs to be registered with the client to be allowed to participate in bidding as a labour-based contractor. Here we present a set of requirements that are often necessary to fulfil but remember, depending on what kind of contracts are offered, different registration requirements might apply. Ask the client what applies to a particular contract you are planning to bid for.

Standard requirements:

- your company should be licensed according to national and local regulations
- a description of previous relevant experience (a record of similar work carried out recently)
- proof that you and your supervisory staff have knowledge of how to apply labour-based methods when maintaining or re-gravelling a road (training certificates)
- resources, such as haulage and compaction equipment for a regravelling contract, must be available.

## RESPOND TO TENDER INVITATION

Once you have been invited to fill out a tender for a certain job, you need to be fully aware of what the requirements are. The first action to take undertake when responding to a tender invitation is to:

- confirm your interest.

You are often expected to indicate your intention to bid for a job (however, only if explicitly mentioned in the tender documents). After you have decided to bid for a project, write a confirmation letter to the client stating your interest in the job and indicating your intention to tender for it.

## STUDY THE TENDER DOCUMENTS

Read the tender documents carefully, including all documents that are referred to, such as standard drawings and specifications. Make sure you fully understand what is expected of you when implementing the contract, concerning quality of output, your rights and your obligations. Also remember to check all drawings and specifications for omissions or mistakes.

If anything in the documents is unclear, you must seek assistance. Do not hesitate to ask the client for clarification of unclear points as it is in the interest of both the client and you to avoid future disputes. For some questions you may not want to consult the client, in which case ask for advice from your contractors' association.

The conditions of contract describe the terms and conditions under which the work is to be done. They impose strict legal and financial obligations on the parties to the contract, i.e. you and the client. They provide very important indications of how the job is likely to run. The risks involved must be taken into account when deciding whether to bid for this job or not, and when pricing your work. Some contracts are not worth bidding for at all, and for others you need to add a very high risk allowance to make it worthwhile. You cannot measure risk unless you know your rights and responsibilities according to the contract.

Go through all the points in the conditions of contract and consider what the consequences will or could be for your work. If you foresee problems, remember to take them into consideration when pricing. The most important points to always look for in the conditions of contract are:

- contract period, including the starting and finishing dates (example of questions to ask yourself: when do I have access to the site?; do I need extra resources to meet the finishing date?)
- method of payment, (for example, what is the procedure for interim payments?; how long after submission should the client pay?; is it likely that I will receive payment on time?)
- retention money, (for example, what is the percentage deducted, and when should it be paid back?)
- payment for materials on site, (for example, what percentage of the value is paid?; what are the conditions?)
- payment for extra work and variations, (for example, what are the conditions for payment for extra work?; what rates are used for variations: from the bill of quantities or decided by the client, or both of you together?)
- price fluctuations, (for example, is there a procedure to compensate for severe inflation?; if not, include an allowance in your bidding price)
- insurance and indemnities, (for example what is the cost of the insurances needed according to the contract?; how are you compensated for your costs?)
- liquidated damages, (for example, how much to pay and under what conditions?; are they reasonable compensation for income lost?; is there a high risk that you will have to pay liquidated damages?)
- extension of time, (for example, are there conditions stated in the contract concerning when you should be allowed

extension of time?; is the length of the extension set by the client or decided by the two of you together?)
- termination of contract, (for example, what are the conditions for termination by the client?; what are the conditions for termination by you?; have you had experience with this client before, and are you likely to run into problems that can lead to termination?)
- arbitration, (for example, who appoints the arbitrator?; what are the steps that must be taken before the case goes to arbitration?)

## SITE VISIT

This is the only way to find out all that you need to know about the site conditions and how they influence your work. Check the conditions under which you will have to carry out each activity and record in your notebook the most important features you discover. To make sure you do not have to revisit the site it is advisable to prepare a list of important issues to be checked.

Make sure the following points are covered when you undertake your site visit. Prepare a strip map of the road plus sketches of important details; take all relevant measurements such as existing cross-section, camber and ditch size; ask local people about weather conditions, soils and labour availability and write down all the features you think might influence the work.

- investigate possible locations for site camp(s), including the client's existing facilities. Inquire about the potential costs of renting a piece of land to establish a camp.
- visit the designated quarries and assess
  - the quality and variability of the material,
  - plans for exploiting the quarry (prepare an outline sketch of the quarry with all important features (for example, where to place overburden, where to stockpile gravel) and include the volume of material to be excavated and hauled
  - the environmental requirements (could there be any soil erosion caused by the work and/or does the quarry need to be filled with overburden at the end of the job?)
  - the condition of the haul road and its maintenance/rehabilitation requirements
- notify the client of any tender document omissions you discover in order to obtain clarifications (for example work standards which are not clearly defined, activities which are described vaguely or are missing).

## COST ESTIMATING

Now it is time to start the 'desk work', i.e. the calculations. However, before you can start pricing each activity you need to 'plan' the work to be carried out in your mind. Here is a good work order to follow.

1. Decide on an overall strategy on how to organize and carry out the work, including the general work method to be used.
2. List all the activities you plan to undertake in the order in which they are to be carried out. Remember also to list all preparatory and support activities (e.g. transport and establishment of site camp, watchmen services, tool maintenance, etc.)
3. For each activity estimate the resources required and calculate the time necessary to complete the task. Your calculations should be based on experience from previous projects (if it is one of your first contracts, you have to assess carefully the available standard rates; what can I achieve?)
4. Decide on site organization arrangements (for example, the number of supervisors, work gangs, responsibilities, support, etc.)
5. Draw up a preliminary work programme.
6. Prepare a schedule of manpower, equipment and materials requirements matching the initial work programme.
7. check the availability of the required manpower, equipment and materials
   a.) within your company,
   b.) available from other companies, organizations or suppliers (hire or purchase), and
   c.) if you need to procure new equipment, check the *realistic* procurement period.
8. Check your work plan against the availability of resources and correct it if necessary.
9. Identify critical activities or events to assist you later in your management decisions (for example, you have placed an order for one tipper truck assuming it will be delivered one week before it is needed at the gravelling site. Delays in the delivery could mean serious problems for you, as you would not be able to achieve the planned output when hauling gravel. Therefore you should, well in advance, come up with alternative plans, such as where can I hire a similar truck at short notice.)

After planning the work, following points 1–9 above, we can now start pricing the activities. In order to prepare a proper cost estimate (pricing) you need first of all to be well aware of all current prices of materials, equipment and labour. But it is equally important to know your overheads (indirect costs). Remember that both the indirect and the direct costs change frequently as prices and service charges change. So you cannot afford to base your calculations on old information. Calculations of rates and overheads need to be reviewed on a regular basis to make sure your calculated price is as correct as possible.

When pricing your bid, the individual calculations are not difficult. However, when pricing you need to make every effort to ensure that *all* your costs are included. So although the calculations are simple, there are so many of them that the main risk you run is to forget to include some of your costs. A good way to help you take all costs into account is to follow a system when pricing. In our system we have divided them into two main groups, direct costs (linked directly to the activities of the job) and indirect costs (the other costs of your company that need to be recovered from projects). For details, please consult IYCB Handbook 1 'Pricing and Bidding'.

The checklist below provides an outline of cost items required.

*Direct Costs*

- labour
- equipment (plus hand tools)
- material
- transportation of material and equipment

*Indirect Costs*

- Preliminaries
  - site camp facilities
  - insurance
  - bonds
  - road signs, safety measures
  - general transport, support or standby equipment
  - tendering costs
  - accommodation
  - salaries, allowances and expenses for supervisory staff
  - hire of support equipment
  - fee to get access to quarry, or other royalties
- Risks

Here are some of the major risks to consider:

- bad weather
- work to be redone
- delayed payment
- carelessness by employees
- rising prices

- Company costs
  - company facilities, i.e. offices, stores, workshop (a proportion of the costs to be covered by each contract)
  - interest on loans, etc. (for example, a bank loan to procure equipment or facilities)
  - depreciation or replacement of equipment and/or facilities (it can be part of the preliminaries, or included in the equipment cost when calculating the direct costs. Most important, make sure it is not forgotten).
  - general expenses (stationery for administration, electricity bill for office etc.)
  - cost of training of staff
  - book-keeping, accountants' and auditors' fees
  - protective clothing (if it can be used on many sites).

*Profit*

This must be based on your assessment of the current situation, concerning both your own company (for example, are staff and equipment under-utilized?) and the overall market (how competitive is the market?).

*Contingencies*

An amount (or a percentage) is often set aside to cover contingencies. It is an allowance for unforeseen costs resulting from activities *not* included in the contract. A contingency post helps both the client and the contractor by offering an opportunity to cope with unforeseen problems and to enhance the value of the job by improving specifications or adding minor items.

If a contingency item is not included in the bill, you can, if you have discovered the item during bid preparation, add a contingency item to the bill. However, before doing so, consult the client to make him or her aware that this item is over and above the price according to the contract documentation. To facilitate matters for the client when comparing different bids, make sure you also present, in your bid, a cost for completion of the works according to the current documentation, i.e. without the contingencies.

*Co-ordination*

After your work plan is ready and the costs are estimated, consider, in detail, how you can fit the work in with other commitments. There might be costs to save by co-ordinating work on this contract with similar assignments on another contract. Check whether certain resources (labour, equipment) could be combined or utilized in a more economical way. For example, the daily hire rate is often cheaper if you hire over a longer period of time. So it might be more economical for you to co-ordinate two different contracts so that the same activity on one contract can follow the other.

Finally, after you have done all the calculations, review the work programme and cost estimates, check for errors and possible improvements and finalize the tender.

## FINANCING

There are two main aspects to financing a construction contract; cash-flow and profit. Even if you ensure that your prices allow a reasonable profit margin, you could be in trouble if you fail to calculate your cash needs throughout the project. To calculate your cash flow properly, you need to know the financial resources required at any time during the contract. The required amount, at a specific time, must be compared with the likely cash availability at that same time. For example, at the beginning of a regravelling contract you need to start up some preparatory activities and procure essential goods immediately. There will, of course, be a delay before you start to receive payments from the client so make sure you also have enough cash to pay your labourers during this period. In some cases you might have access to a mobilization advance from the client, but that should not be a reason for not preparing a cash-flow estimate.

Concerning financial matters, there are a number of issues which you, as a contractor must consider. Here is a check-list of the most important points.

- Review existing cash resources and commitments to other work.
- Draw up a forecast of outgoing cash for each month of the contract.
- Estimate realistically the cash income from work completion certificates for each month of the contract.
- Estimate additional cash needs (if outgoing cash is larger than the income) for each month of the contract (best and worst cases).
- Investigate, well in advance, possible sources of finance to fill any expected gap.
- Make provisional arrangements for bonds, insurance and other required services.

## TENDER SUBMISSION

The following guidelines may help you when preparing and submitting a tender:

- Prepare the tender documents based on similar work done previously. Make sure you have all the data available and you know exactly where the problems were with the last contract, and where you made a good profit.
- Check carefully that the tender meets all the requirements and that nothing has been overlooked. Put extra effort into checking that all the arithmetic is correct, because this is often a major cause of losing a contract, or getting a contract with prices that results in a loss.
- Make arrangements for secure delivery of the tender by the submission deadline, and that the submission requirements (such as the number of copies, or the use of a sealed envelope) are met exactly.
- Remember that preparing a bid always takes more time than you expect, so start your preparations early and leave plenty of time for checking your calculations.

After the tenders have been submitted attend the tender opening session and note the tender details of your competitors.

## CONTRACT NEGOTIATIONS

Before a contract is finalized, there might be a need to negotiate details which are unclear.

- Clarify any matters raised by the client and correct any errors identified.
- Negotiate on difficult issues, but remember that to negotiate successfully you must know all the relevant facts and be able to foresee the consequences of any changes to your tender submission. The only way to make sure you are in such a position is to prepare your bid thoroughly.

## MOBILIZATION

Once you have been awarded a contract you need to start preparing for the work. If these activities are not carried out in time you will not be able to start the contract activities on time as agreed in the programme. These activities are also called mobilization activities and consist of the following:

- Confirm all financial arrangements previously made provisionally.

- Appoint and brief the supervisory staff about the new contract and make early arrangements for the deployment of the field staff.
- Prepare the labour recruitment (inform local leaders and authorities, inform job seekers, liaise with the labour office and unions).
- Set up site camps.
- Introduce planning, reporting and control procedures for all relevant staff.
- Procure hand tools, equipment and materials, and transport them to the site.
- Make quarry arrangements (compensation of landowners, setting out of quarry, and so on).
- Make arrangements for servicing and repair of vehicles and equipment.
- Train staff and labourers to enable them to carry out their tasks.
- Brief your supervisors about work details.
- Set up arrangements for paying labourers and other items.
- Arrange for adequate safety and first aid measures.

## WORK MANAGEMENT

When the contract implementation starts it is necessary for you to monitor the work carefully throughout the contract period. There are a number of management activities that need to be carried out:

- Review the contract work programme and make any necessary changes with the agreement of the client or his representative.
- Set up daily planning and reporting arrangements and make sure you get reports in time to analyse the data.
- Review the plan for each day's work at least the day before.
- Advise supervisors and workers of their duties the day before, in order to avoid delays each morning.
- Arrange for necessary transport as early as possible.
- Set out daily tasks where possible.
- Monitor performance carefully, where work is issued on a daily basis.
- Monitor the quality of work throughout the day using appropriate measuring aids, such as templates and camberboards.
- Record inputs and outputs each day, and physical progress (for example, km, $m^3$, trips per hauling equipment) and compare with the work programme.
- Check productivities carefully and identify reasons for not achieving the set targets.

- Check your bill of quantities and note work actually carried out. Inform the client immediately where deviations are apparent and, if necessary, put it in writing and inform the client of the resulting additional expenditures.
- Rectify any substandard work immediately and make arrangements to avoid a repetition.
- Monitor the overall programme regularly with the client or his representative and update it if necessary. Inform the client of any deviations from the programme and, if necessary, explain the reasons in writing.
- Adjust the programme and resource requirements if additional work is instructed by the client or his representative. Make sure you get instructions for any additional work in writing. When you receive instructions to undertake activities which are not priced in the contract, agree on a price or a rate with the client before the extra work starts. Remember also to confirm this price in writing to the client. If the contract period is extended because of additional work instructed by the client, the client should be informed accordingly in writing to obtain his or her confirmation.
- Monitor the actual cash-flow regularly against the forecast plan, and take any necessary action.
- Make timely arrangements for payments to labourers and staff and, if necessary, make any security arrangements that might be required for the pay days.

## PAYMENT CERTIFICATES

After a certain amount of work has been carried out, or after a certain period (often every month), you and the client (or his representative) will inspect and measure the work done. If the work has been satisfactorily carried out a payment certificate can then be prepared.

- prepare certificates based on the achieved output as recorded in the site reports and as inspected by the client.
- Agree with client, and let the client sign the certificates.

## MONITORING PERFORMANCE

To ensure that your contract is carried out as efficiently as possible, i.e. make as much profit as possible, you must carefully measure the work performance and compare it with the plans prepared when bidding. This monitoring enables you to improve unsatisfactory performance at an early stage (before a substantial loss has been accumulated). Your records should be kept in such a way that they can be of help when planning future contracts. In particular, you should re-calculate prices after the work is done and try to identify areas where you need to make improvements or where your future prices must be changed.

To be fully informed about the project and to be able to manage it properly you must monitor:

- physical progress
- resources used
- productivity achieved
- supervisory performance
- fuel consumption
- availability and utilization of equipment
- reasons for down-time of equipment
- costs
- problems encountered and proposed solutions.

## TRAINING AND MANPOWER DEVELOPMENT

Training makes it possible, or easier, for your staff to adapt quickly to new work methods, and to achieve increased work output. Although it might appear costly in the short run, training often pays for itself manyfold in the longer run. Training, in this context, does not necessarily mean a formal course where members of your staff are away from site for several weeks. It can also mean on-the-job training that causes less interruption to your work plan.

As you are well aware, your main means of production on a labour-based road contract are the labourers, so it is important for you to earn a reputation as a good employer. One way of earning this reputation is to give the labourers a chance to develop their skills.

To achieve this you should:

- monitor personnel capabilities, development and the resulting training-needs for enhanced performance or new work methods
- arrange for adequate initial, upgrading or refresher training, where possible on-the-job
- analyse the overall work situation so that you can prepare for training programmes to meet the work requirements
- instruct and coach your supervisors on every possible occasion; organize regular meetings where you allow staff to propose work improvements or changes; counsel staff wherever possible and give them the opportunity to approach you with their problems

- give your staff possibilities to undertake more advanced work gradually. The key is often to trust your staff and their capacity, but remember not to rush. Not being able to fulfil what your chief expects can have long-term discouraging effects.

### DISPUTES AND ARBITRATION

There will always be differences between the contractor's interpretation of the contract documents and the way the client wants the job to be done. Your aim as a contractor should be on the one hand to satisfy the client, and on the other hand to make sure you can run your business profitably. Often it will be necessary to work out a compromise with the client.

Some general advice on how to handle disputes:

- Always inform the client at an early stage of any problem or disagreement.
- Make every effort to resolve any dispute within the framework of the contract.
- Record all necessary data representing the true development of a dispute.
- Always make sure that instructions from the client which deviate from the contract documents are given in writing, and remember to confirm in writing oral agreements reached during discussions or meetings. You can prepare minutes of the meetings and let the client sign these minutes.

## Pricing for Routine Maintenance Contracts

When routine maintenance contracts are issued based on itemized and measured activities, calculation of prices is done in the same way as for any normal building or construction contract.

Where routine maintenance contracts are issued on the basis of standards to be achieved and maintained over a certain period of time, the required workload has to be assessed by the contractor and the items priced accordingly.

The procedure for estimating costs for routine maintenance activities is very much the same as explained in the IYCB Handbook 'Pricing and Bidding'. Without going through all the pricing details we shall, in this chapter, work out a sample contract to illustrate how the estimating is done for a routine maintenance contract, assuming that the pricing details are already known to you.

For our pricing exercise we shall refer to the routine maintenance contract example mentioned in Chapter 7.

Example:

*Brief contract description*

| | |
|---|---|
| Road name: | Bonoki–Maruba |
| Road length: | 34.000km |
| Carriageway: | 5.50m wide, gravel layer with standard thickness of 15cm (compacted), crossfall 5% |
| Cross-section: | |

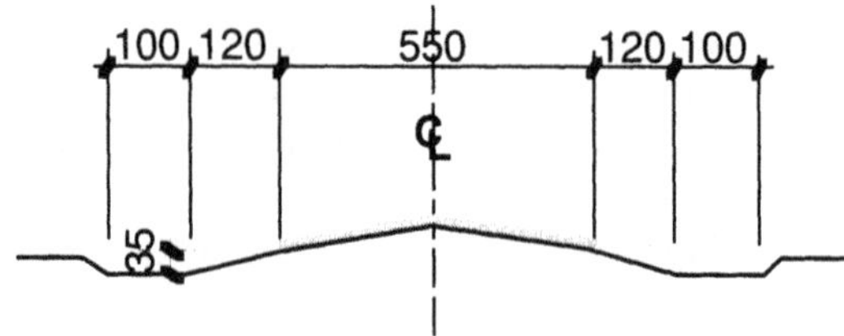


| | |
|---|---|
| Climate: | little rain in March and April (approximately 800mm p.a.) |
| Labour available: | average of 220 habitants per km$^2$ |
| General condition: | good condition, no major faults, drains well-established, uniform thickness of gravel layer |
| Contract period: | 3 months (1.1.96 to 31.3.96) |

*Bill of Quantities*

| Item | Description | Unit | Quantity | Rate | Amount |
|---|---|---|---|---|---|
| 1 | Inspection and removal of obstructions and debris over the entire road length | day | 24 | | |
| 2 | Clear silt and debris from culvert, inlet and outlet ditch, and dispose of this material safely inside or outside the road reserve | No. | 102 | | |
| 3 | Clean the side drains to the standard cross-section removing all soil, vegetation and other debris and disposing of safely inside or outside the road reserve | m | 14 500 | | |
| 4 | Repair scour checks to standard dimensions using stones or wooden sticks | No. | 56 | | |

| Item | Description | Unit | Quantity | Rate | Amount |
|---|---|---|---|---|---|
| 5 | Clean the mitre drains to the standard cross-section removing all soil, vegetation and other debris and disposing of safely inside or outside the road reserve | m | 1850 | | |
| 6 | Fill pot-holes with approved material and compact well | No. | 120 | | |
| 7 | Grub carriageway edge and remove roots and spoil well clear of the carriageway on to the road service | m | 9200 | | |
| 8 | Repair shoulder and slope erosion | m | 250 | | |
| 9 | Cut grass to a maximum height of 50mm above ground level on shoulders, ditch bottoms and side slopes | $m^2$ | 22 050 | | |
| 10 | Clear bush and remove roots and stumps from side ditch, shoulder, slope and road reserve 2m from outer side of ditch | $m^2$ | 4400 | | |
| | Total | | | | |
| 11 | Miscellaneous, 5% of total | | | | |
| | Total, incl. Miscellaneous | | | | |

Note: point 11. Miscellaneous was included by the client as a 5% addition to the total. Although we have to include item 11 in our bill, it is useful to separate this from the other items. Separation makes it easier to compare costs on this contract with other contracts, and to extract productivity figures to use when preparing other bids.

Based on the volume of work as described in the bill of quantities our first assumptions are as follows:

*Approximate work-load*
600 work-days

*Assumed organization and tools*
Two gangs, each with 8 labourers and one gangleader, gangs would work, on an average, 3 days per week, gangleaders to be provided with bicycles. Labourers to be provided with standard hand tool set of each:
1 hoe (total 16 = 128 NU)
1 shovel (total 16 = 192 NU)
1 rake (total 16 = 144 NU)
1 grass slasher (total 16 = 80 NU)
1 bush knife (total 16 = 96 NU)
**total investment cost = 640 NU**

in addition, for each gang:
3 wheelbarrows (total 6 = 336 NU)
1 long-handled spade (total 2 = 24 NU)
1 long-handled trowel (total 2 = 22 NU)
2 pick axes (total 4 = 48 NU)
1 axe (total 2 = 32 NU)
**total investment cost = 462 NU**

in addition for each gangleader
1 bicycle (total 2 = 420 NU)
1 spirit-level (total 2 = 12 NU)
1 set of boning rods (total 2 = 6 NU)
1 ditch template (total 2 = 6 NU)
1 straight edge (total 2 = 4 NU)
1 line level (total 2 = 6 NU)
strings (total = 6 NU)
**total investment cost = 460 NU**

*Transport*
1 pickup, approximately 60km per site visit, total 12 visits = total 720 km plus 40% for procurement of goods, visits to client, etc. = total 1008 km at 0.35 NU per km = 352.80 NU

*Daily labour wage rate*
1.50 NU

*Daily gangleader rate*
3.00 NU

As a first step we will need to calculate the direct project costs using the IYCB standard calculation charts. To calculate the different BQ items we have to rely on our past experience and our productivity tables (task rates).

Tools are, together with the labourers, our main means of production and we are therefore including them as part of the direct costs rather than as an indirect cost element. For the tools we assume that we must write off the standard hand tool set after one year, or 220 working days. With 16 labourers we can produce 3520 days of work. The total cost for the hand tools was 640 NU. We are therefore getting an average hand tool rate of 0.18 NU per work-day.

The major assumptions are stated in the following direct project costs chart:

| Direct Project Cost Chart | | | | | | | | |
|---|---|---|---|---|---|---|---|---|
| List of quantities | | | | Direct project costs (NU) | | | | |
| Item No. | Description | Unit | Quantity | Labour | Plant tools | Material | Transport | Total |
| 1 | Inspection and removal of obstructions and debris | day | 24 | 36.00 | 4.32 | | | 40.32 |
| 2 | Clean culverts | No. | 102 | 76.50 | 9.18 | | | 85.68 |
| 3 | Clean side drains | m | 14 500 | 435.00 | 52.20 | | | 487.20 |
| 4 | Repair scour checks | No. | 56 | 16.50 | 1.98 | | | 18.48 |
| 5 | Clean mitre drains | m | 1850 | 46.50 | 5.58 | | | 52.08 |
| 6 | Fill pot-holes | No. | 120 | 9.00 | 1.08 | | | 10.08 |
| 7 | Grub carriageway edge | m | 9200 | 55.50 | 6.66 | | | 62.16 |
| 8 | Repair shoulder and slope | m | 250 | 7.50 | 0.90 | | | 8.40 |
| 9 | Cut grass | $m^2$ | 22 050 | 94.50 | 11.34 | | | 105.84 |
| 10 | Clear bush | $m^2$ | 4400 | 33.00 | 3.96 | | | 36.96 |
| | Total direct project cost | | | | | | | 907.20 |
| 11 | Miscellaneous (if required) (5%) | | | | | | | 45.36 |
| | Final total direct project cost | | | | | | | 952.56 |

## CALCULATIONS

*Item 1: Inspection and removal of debris*

Assumption: 24 days indicated in the bill of quantities

Cost: labour 24 wd × 1.50 NU = 36.00 NU

tools 24 wd × 0.18 NU = 4.32 NU

*Item 2: Clean culverts*

Assumption: Productivity rate = 2 culv./wd;
total work-days: 102m/2No./wd = 51 wd
Cost: labour 51 wd × 1.50 NU = 76.50 NU
tools 51 wd × 0.18 NU = 9.18 NU

*Item 3: Clean side drains*

Assumption: Productivity rate = 50m drains/wd;
total work-days: 14 500m/50m/wd = 290 wd
Cost: labour 290 wd × 1.50 NU = 435.00 NU
tools 290 wd × 0.18 NU = 52.20 NU

*Item 4: Repair scour checks*

Assumption: Productivity rate = 5 No./wd;
total work-days: 56/5 No./wd = 11wd
Cost: labour 11 wd × 1.50 NU = 16.50 NU
tools 11 wd × 0.18 NU = 1.98 NU

*Item 5: Clean mitre drains*

Assumption: Productivity rate = 60m/wd;
total work-days: 1850/60m/wd = 31 wd
Cost: labour 31 wd × 1.50 NU = 46.50 NU
tools 31 wd × 0.18 NU = 5.58 NU

*Item 6: Fill pot-holes*

Assumption: Productivity rate = 20 No./wd;
total work-days: 120 No./20 No./wd = 6 wd
Cost: labour 6 wd × 1.50 NU = 9.00 NU
tools 6 wd × 0.18 NU = 1.08 NU

*Item 7: Grub carrigeway edge*

Assumption: Productivity rate = 250m/wd;
total work-days: 9200m/250m/wd = 37 wd
Cost: labour 37 wd × 1.50 NU = 55.50 NU
tools 37 wd × 0.18 NU = 6.66 NU

*Item 8: Repair shoulder and slope*

Assumption: Productivity rate = 50m/wd;
total work-days: 250m/50m/wd = 5 wd
Cost: labour 5 wd × 1.50 NU = 7.50 NU
tools 5 wd × 0.18 NU = 0.90 NU

*Item 9: Cut grass*

Assumption: Productivity rate = 350m$^2$/wd;
total work-days: 22 050m$^2$/350m$^2$/wd = 63 wd
Cost: labour 63 wd × 1.50 NU = 95.50 NU
tools 63 wd × 0.18 NU = 11.34 NU

*Item 10: Clear bush*

Assumption: Productivity rate = 200m²/wd;
total work-days: 4400m²/200m²/wd = 22 wd

Cost: labour 22 wd × 1.50 NU = 33.00 NU
tools 22 wd × 0.18 NU = 3.96 NU

You can see that there are no costs entered in the material and transport columns of the Direct Project Cost Chart on page 241. None of the items require materials to be brought to the site and the transport cost for site visits, as calculated on page 240, is not linked to individual items. The transport cost will therefore be included with the preliminaries together with other costs related to general supervision of the works.

In the next step we have to calculate the indirect costs. It is best to add these costs as a percentage to each item. The total indirect costs for this contract are:

*Preliminaries*

Supervision: 2 gangleaders for 12 weeks @ 3 working days per week = total 72 work-days

| Item | Description | Cost NU |
|---|---|---|
| P1 | Supervision:<br>2 Gangleaders for 12 weeks @ 3 days = 72 days @ 3.00 NU = 216 NU<br>gangleader's equipment total 460 NU over 2 years, or 440 days; rate per day = 460 NU/440 days = 1.05 NU/day, for 36 days = 37.80 NU<br>gang tool set total 462 NU over 2 years or 440 days; rate per day = 1.05, for 36 days = 37.80 NU | 291.60 |
| P2 | Contractor's fully comprehensive insurance: | 100.00 |
| P3 | Transport for site visits (for cost calculation, see page 240) | 352.80 |
| | Total Preliminary Costs | 744.40 |

*Risk Allowance*

We are adding the risk allowance as a percentage to the total direct costs. For routine maintenance work there is a danger that the estimates are always inaccurate to some degree, as initial measurements can never be very exact. We are therefore assuming a risk allowance rate of 6% of the total direct costs (952.56 NU) which gives us 57.15 NU.

Total risk allowance = 57.15 NU

*Company Costs*

To calculate the company costs we have to know the total workload of the company at the time of this particular contract. In our case we have two other contracts of similar size and we therefore share the company costs equally between the three contracts. The administration of our small company is carried out by your wife on a part-time basis.

| | |
|---|---|
| Director's salary | 2200 NU |
| Administration | 120 NU |
| Bookkeeper, Auditor | 80 NU |
| Interest on bank loan | 120 NU |
| **Total for one year** | **2520 NU** |
| monthly average costs (2520/12) | 210 NU |
| **For this contract** | |
| (210 × 3 months/3 contracts) | **210 NU** |

*The total indirect costs are*

| | |
|---|---|
| Preliminary costs | 745 NU |
| Risk allowance | 57 NU |
| Company costs | 210 NU |
| **Total** | **1012 NU** |

The total direct costs are 952.56 NU and therefore the indirect costs are approximately an additional 106%. This appears to be a lot, but is quite normal for a routine maintenance contract in many countries. Labour-based maintenance needs a lot of supervision and average daily labour rates are often very low.

Before we add up all the figures we can assume the size of the expected profit. As there is stiff competition we limit the profit margin to 8% of the direct costs. This will be added as a percentage to each contract item.

Our final cost chart now looks like this:

| Item No. | Description | Direct costs NU | Indirect costs NU | Profit 8% NU | Total NU |
|---|---|---|---|---|---|
| 1 | Inspection and removal of obstructions | 40.32 | 42.74 | 3.22 | 86.28 |
| 2 | Clean culverts | 85.68 | 90.82 | 6.85 | 183.35 |
| 3 | Clean side drains | 487.20 | 516.43 | 38.98 | 1042.61 |
| 4 | Repair scour checks | 18.48 | 19.59 | 1.48 | 39.55 |
| 5 | Clean mitre drains | 52.08 | 55.20 | 4.17 | 111.45 |
| 6 | Fill pot-holes | 10.08 | 10.68 | 0.81 | 21.57 |
| 7 | Grub carriageway edge | 62.16 | 65.89 | 4.95 | 133.00 |
| 8 | Repair shoulder | 8.40 | 8.90 | 0.67 | 17.97 |
| 9 | Cut grass | 105.84 | 112.19 | 8.46 | 226.49 |
| 10 | Clear bushes | 36.96 | 39.18 | 2.96 | 79.10 |
| | Total | 907.20 | 961.62 | 72.55 | 1941.37 |
| 11 | Miscellaneous | 45.36 | 48.08 | 3.90 | 97.34 |
| | Total, incl Miscellaneous | 952.56 | 1009.70 | 76.45 | 2038.71 |

For the last step we now have to calculate the unit rates and fill out the BQ. The process to follow can be illustrated by an example, Item 3, 'Clear side drains'. First you divide the total cost from the final cost chart (on the previous page) with the quantity indicated in the BQ (NU 1042.61 / 14 500 m = 0.07190 NU/m). You round off the figure for the rate (usually not more than 3 decimal figures) to 0.072 NU/m. You take this rate (0.072 NU/m) and multiply it with the quantity (14 500m) and arrive at the amount to put in your BQ (0.072 NU/m x 14 500m = 1044 NU).

The BQ with our final offer looks like this:

*Bill of Quantities*

| Item | Description | Unit | Quantity | Rate | Amount |
|---|---|---|---|---|---|
| 1 | Inspection and removal of obstructions and debris over the entire road length | day | 24 | 3.60 | 86.40 |
| 2 | Clear silt and debris from culverts, inlet and outlet ditches, and dispose of this material safely inside or outside the road reserve | No. | 102 | 1.80 | 183.60 |
| 3 | Clean the side drains to the standard cross-section removing all soil, vegetation and other debris and disposing of safely inside or outside the road reserve. | m | 14 500 | 0.072 | 1044.00 |
| 4 | Repair scour checks to standard dimensions using stones or wooden sticks. | No. | 56 | 0.71 | 39.80 |
| 5 | Clean the mitre drains to the standard cross-section removing all soil, vegetation and other debris and disposing of safely inside or outside the road reserve. | m | 1850 | 0.060 | 111.00 |
| 6 | Fill pot-holes with approved material and compact well. | No. | 120 | 0.18 | 21.60 |
| 7 | Grub carriageway edge and remove roots and spoil well clear of the carriageway on to the road service. | m | 9200 | 0.015 | 138.00 |
| 8 | Repair shoulder and slope erosion. | m | 250 | 0.072 | 18.00 |
| 9 | Cut grass to a maximum height of 50mm above ground level on shoulders, ditch bottoms and side slopes. | $m^2$ | 22 050 | 0.010 | 220.50 |
| 10 | Clear bushes and remove roots and stumps from side ditch, shoulder, slope and road reserve 2m from outer side of ditch. | $m^2$ | 4400 | 0.018 | 79.20 |
| | Total | | | | 1942.10 |
| 11 | Miscellaneous, 5% of total | | | | 97.10 |
| | Total, incl Miscellaneous | | | | 2039.20 |

The price for the contract to be offered to the client is thus: 2039.20 NU

# Pricing for Regravelling Contracts

Regravelling contracts usually take the same form as normal construction contracts and the procedure to estimate the costs for regravelling activities is the same as explained in IYCB Handbook 1, 'Pricing and Bidding'. Without going through all the pricing details here, we will calculate a sample contract to illustrate how the estimating is done for a regravelling contract, assuming that the pricing details are already known to you.

For our pricing exercise we refer to the regravelling contract example mentioned in Chapter 8.

Example:

*Brief contract description*

| | |
|---|---|
| Road name: | Village A–Village C |
| Road length: | 15.000km |
| Carriageway: | 4.00m wide, gravel layer with standard thickness of 15cm (compacted), crossfall 5% |
| Cross-section: | |

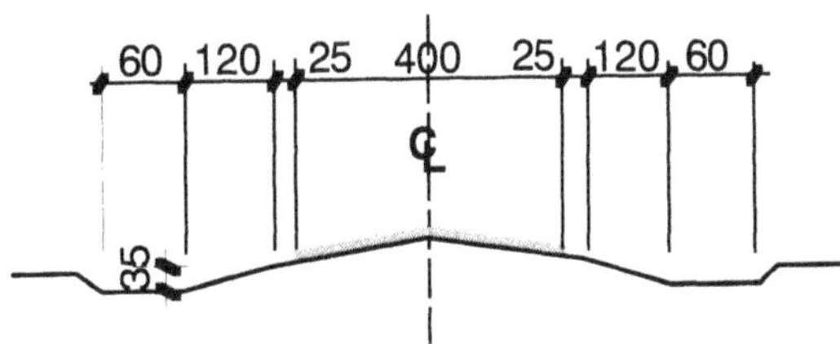


| | |
|---|---|
| Climate: | rain in March/April and November (approx. 1400mm p.a.) |
| Labour availability: | average of 300 habitants per $km^2$ |
| General condition: | good condition, no major faults, drains well established, carriageway slightly deformed (ruts, pot-holes) |
| Contract period: | 4 months |

*Bill of Quantities*

| Item | Description | Unit | Quantity | Rate | Amount |
|---|---|---|---|---|---|
| 1 | Establishment | lump sum | | | |
| 2 | Improvement of quarry access roads including maintenance throughout the contract period | m | 12 000 | | |
| 3 | Quarry preparation; consisting of bush clearing, grass cutting, grubbing and clearing well of the quarry | $m^2$ | 5800 | | |
| 4 | Excavation of overburden including loading, hauling and stocking within 100m | $m^3$ | 1820 | | |
| 5 | Reshaping of road, consisting of re-establishment of carriageway crossfall, reshaping of shoulders and slopes and cleaning of ditches, mitre drains and culverts | m | 1500 | | |
| 6A | Excavation of *in situ* gravel and stockpiling ready for loading (hard) | $m^3$ | 4180 | | |
| 6B | Excavation of *in situ* gravel and stockpiling ready for loading (very hard) | $m^3$ | 7220 | | |
| 6C | Excavation of *in situ* gravel and stockpiling ready for loading (soft) | $m^3$ | Prov. | | |
| 7 | Loading of loose gravel on to trailer or lorry | $m^3$ | 11 400 | | |
| 8A | Hauling of gravel to site; average haul distance 3.500km | $m^3$ | 4180 | | |
| 8B | Hauling of gravel to site; average haul distance 7.000km | $m^3$ | 4940 | | |
| 8C | Hauling of gravel to site; average haul distance 1.500km | $m^3$ | 2280 | | |
| 9 | Off-loading and spreading of gravel material to the required thickness | $m^3$ | 11 400 | | |
| 10 | Watering and compacting of gravel layer by vibrating roller to the required camber at OMC ±2% in layers of no more than 150mm compacted in carriageway | $m^2$ | 60 000 | | |
| | TOTAL | | | | |

**General Situation:** we have made a site visit and have produced a strip map with the following features and data:

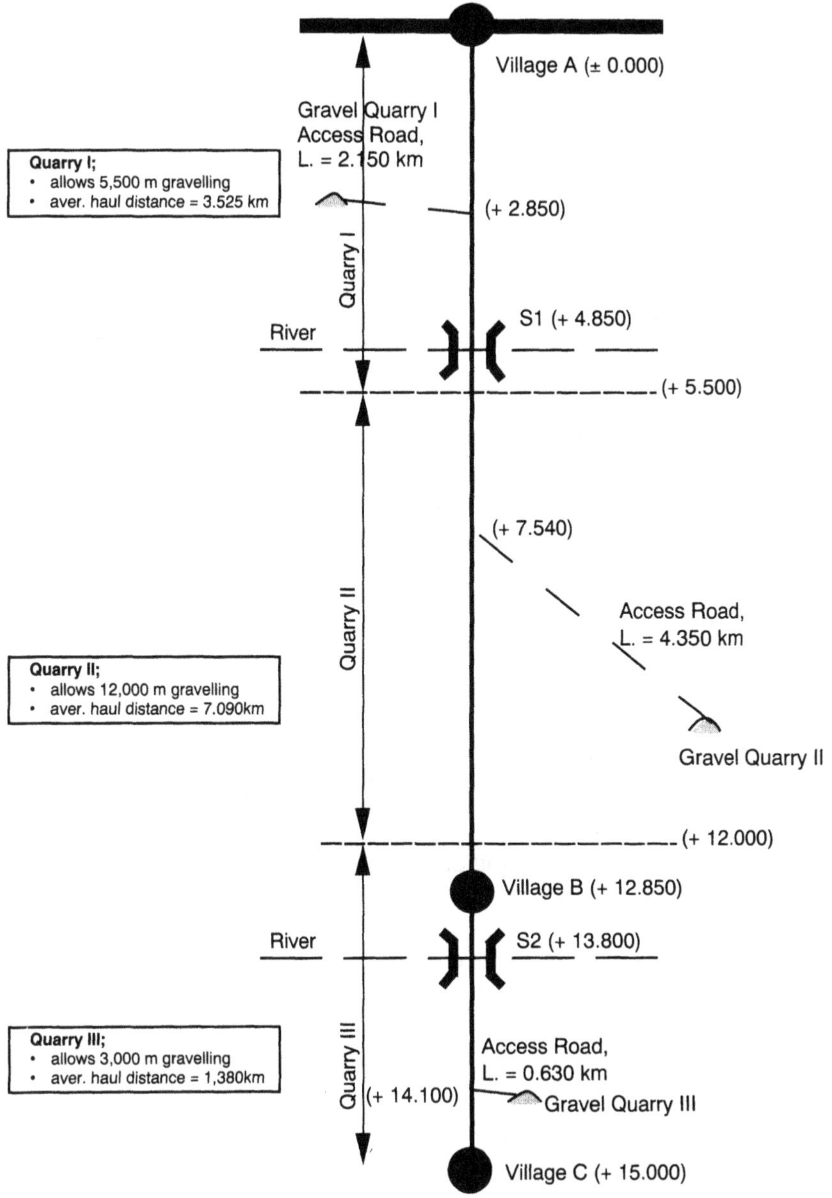


Based on this assessment we have prepared preliminary assumptions and calculations on the required work organization and productivities to complete the work in the required contract period of 4 months.

Please note that there is a major difference between your strip map and the Bill of Quantities concerning the length of the quarry access roads (12 km compared to 7 km). Start by checking your own figures. If your figures seem right, inform the client. In the following calculations the 'assumptions and Initial calculations' section (pages 243–247) are based on the strip map while the detailed cost calculations on pages 250–253 rely on the provided BQ. Check the two against each other to make sure that all your costs are covered.

*Assumptions and Initial Calculations*
Gravel:
The contract specifies a surface width of 400cm with a compacted gravel layer of 15cm. In order to get the necessary thickness of loose material we add 25% to 15cm which gives us 19cm (rounded up). Per metre road length we therefore require 0.19 x 4.00 = 0.76m$^3$ of loose gravel.

*Equipment requirements*
The equipment days (in our case lorries) will be based on the quantity of gravel to be transported and the average hauling distance for each section of road:

Quarry I;
Average hauling distance = 3.525km
Amount of gravel to be hauled = 4180m$^3$
Lorry capacity = 4m$^3$
Total lorry loads required 4180/4 = 1045 loads (trips)
Trips per lorry per day (Table 8.13, good haul route) = 12 trips
**Total required lorry-days for Quarry I (1045 / 12) = 87 lorry-days**

Quarry II;
Average hauling distance = 7090km
Amount of gravel to be hauled = 4940m$^3$
Lorry capacity = 4m$^3$
Total lorry loads required (4940/4) = 1235 loads (trips)
Trips per lorry per day (see Table 8.12, average haul route) = 8 trips
**Total required lorry-days for Quarry II (1235/8) = 155 lorry-days**

Quarry III;
Average hauling distance = 1380km
Amount of gravel to be hauled = 2280m$^3$
Lorry capacity = 4m$^3$
Total lorry loads required (2280/4) = 570 loads (trips)
Trips per lorry per day (see Table 8.12, good haul route) = 14 trips
**Total required lorry-days for Quarry III (570/14) = 41 lorry-days**
**Total lorry-days required all quarries (87 + 155 + 41) = 283 lorry-days**

In practice equipment can usually not be fully utilised (100%). There are always minor break-downs, delays in loading and off-loading, poor weather conditions, and so on that do not allow you to produce at a maximum level. Therefore it is necessary to take these delays into account when calculating the time required to complete the work. In our example we assume, based on previous experience, that we will be able to utilize the lorries at an effective rate of 90%.

With an effective utilization rate of 90% we will arrive at 314 lorry-days (283 / 0.90 = 314) needed.

So, in total we have estimated 314 lorry days. The actual time required to haul the gravel to site now depends on the number of lorries we assign for the job. In our case we had initially thought of using only the two lorries we recently bought. However, we would then require about 157 days, or over 7 months (7.2), just to transport the gravel.

The contract specifies a total time for completing the works of 4 months and we have to set aside some time to carry out all the preparatory work and to excavate and stockpile sufficient gravel. These activities we plan to complete in one month. We are therefore left with 3 months for the actual gravelling job. The average working days per month are estimated to be 22 which gives us 66 working days. The lorry-days required are 314 and if we divide that by 66 working days we need 4.8 lorries to complete the work in time. This means that we require 5 lorries to meet the contract requirements. Using 5 lorries will even allow a slightly lower utilization rate than 90%, facilitating for us to achieve the daily targets. Subsequently we will base our calculations on 66 gravelling days. The number of days taken up for transportation of gravel also influences other activities, like loading and spreading. If transporting the gravel takes 66 days, loading and spreading will of course also take 66 days. However, note that excavation of gravel can normally be done independently from transport of gravel since the gravel can be stockpiled (if space is available for stockpiling).

We have decided to hire three additional lorries for three months from another company to supplement our fleet of two lorries.

In addition we require two pedestrian vibrating rollers and one tractor with a water bowser to fetch the water from the river and to spray it on to the gravel course. This equipment is available in our company. The tractor with a trailer is also used to transport materials and tools when required.

For supervision we also require a pick-up on a part-time basis. Our company is still working on another small building job and we therefore assume that the pick-up will be assigned to the regravelling job for 75% of the time.

An estimate of the total equipment requirements for this job is:

- 2 of our own lorries for 66 working days
  (costs per hour and lorry 28 NU = per 8 hour day = 224 NU)
  total daily costs for 2 lorries 448 NU and for 66 days = 29 568 NU

  - 3 hired lorries for 66 working days
    (hired for a lump-sum amount for 66 working days) = 50 000 NU
  - 2 pedestrian vibrating rollers for 66 working days
    (costs per hour and roller 8 NU = per 8 hour day = 64 NU)
    total daily costs for 2 rollers 128 NU and for 66 days = 8448 NU
  - 1 tractor with water bowser and trailer for 66 working days
    (costs per hour 18 NU = per 8 hour day = 144 NU)
    total daily costs 144 NU and for 66 days = 9504 NU
- 1 pick-up for 75% of 88 working days = 66 working days
  (costs per hour 5 NU = per 8 hour day = 40 NU)
  total daily costs 40 NU and for 66 days = 2640 NU

*Labour Requirements*

The labour requirements for each section (see strip map) are:

Quarry I: From our site visit notes we know that Quarry I is 30 x 50m in size and that the overburden is 30cm as an average.

- length to be gravelled = 5500m (Figure 9.3)
- loose gravel required = length x volume gravel per metre of road 5500 x 0.76 = 4180 m$^3$
- work-days required:
  The work-days required are calculated using the tables of task rates previously presented (Tables 8.8 – 8.11)

Quarry preparation; length access road = 2150m (Figure 9.3)

- access road improvement (Table 8.8) =
  2150m / 60m per work-day = 36wd
- bush clearing (Table 8.8) = 1500m$^2$ / 500m$^2$ p.wd = 3wd
- overburden removal (Table 8.8) =
  450m$^3$ / 3m$^3$ p.wd = 150wd

Road reshaping;

○ reshaping (Table 8.9) = 5500m / 40m p.wd = 138wd

Gravelling:

○ excavation hard gravel (Table 8.10) =
4180m$^3$/1.7m$^3$ p.wd = 2459wd

○ loading to lorry (Table 8.11) =
(4180m$^3$ / 7m$^3$ p.wd) x 1.11 = 662wd

○ off-loading + spreading (Table 8.12) =
(4180m$^3$ / 15m$^3$ p.wd) x 1.11 = 309wd

○ miscellaneous (support, reshaping, etc.) + 10%
(3757 x 0.10) = 376wd

**Total Quarry I** **4133wd**

Due to the reduced utilization rate of the lorries (90%) we also need to consider a lower 'utilization' of labourers carrying out activities directly related to the hauling. In practical terms you would have to allocate labourers to loading as well as off-loading and spreading on the assumption that we would achieve the maximum possible number of trips by the lorries per day. However, a major breakdown affecting one of the lorries would probably make it impossible to achieve the maximum number of trips that we based our calculations on. If the number of trips is reduced the labourers will not be able to work at a maximum level either, i.e. we will have an under-utilization of the labour as well. In order to take this reduction into account we assume the same 'utilization rate' for the labourers as for the lorries – 90%. This gives us a multiplication factor of 1.11 for the work-days (100% / 90% = 1.11). This factor is employed above for the loading and the off-loading activities, while the excavation is not affected since you can normally stockpile gravel.

Quarry II:

- length to be gravelled = 6500m (Figure 9.3)
- loose gravel required = 4940m$^3$ (6500 x 0.76)
- work-days required:

Quarry preparation; length access road = 4350m (Figure 9.3)

○ access road improvement (Table 8.8) =
4350m / 60m p.wd = 73wd

○ bush clearing (Table 8.8) = 3500m$^2$ / 300m$^2$ p.wd = 12wd

○ overburden removal (Table 8.8) =
1050m$^3$ / 3m$^3$ p.wd = 350wd

Road reshaping;

- reshaping (Table 8.9) = 6500m / 40m p.wd = 163wd

Gravelling:

- excavation of oversize gravel (Table 8.10) = $4940m^3$ / $1.2m^3$ p.wd = 4117wd
- loading to lorry (Table 8.11) = ($4940m^3$ / $7m^3$ p.wd) x 1.11 = 783wd
- off-loading and spreading (Table 8.12) = ($4940m^3$ / $15m^3$ p.wd) x 1.11 = 366wd
- miscellaneous (support, reshaping, etc.) + 10% = 586wd

**Total Quarry II** **6450wd**

Quarry III:

- length to be gravelled = 3000m (Figure 9.3)
- loose gravel required = $2280m^3$ (3000 x 0.76)
- work-days required:

Quarry preparation; length access road = 630m (Figure 9.3)

- access road improvement (Table 8.8) = 630m / 60m p.wd = 11wd
- bush clearing (Table 8.8) = $800m^2$ / $300m^2$ p.wd = 3wd
- overburden removal (Table 8.8) = $320m^3$ / $3m^3$ p.wd = 107wd

Road reshaping;

- reshaping (Table 8.9) = 3000m / 40m p.wd = 75wd

Gravelling:

- excavation of oversize gravel (Table 8.10) = $2280m^3$ / $1.2m^3$ p.wd = 1900wd
- loading to lorry (Table 8.11) = $2280m^3$ / $7m^3$ p.wd x 1.11 = 362wd
- off-loading and spreading (Table 8.12) = $2280m^3$ / $15m^3$ p.wd x 1.11 = 169wd
- miscellaneous (support, reshaping, etc.) + 10% = 263wd

Total Quarry III 2890wd

**Total workdays required all quarries** (4133 + 6450 + 2890) **13,473wd**

You now have a very good first indication of the required manpower and equipment. Based on your final decision on how many pieces of equipment and how many labourers you will use for a particular activity a detailed cost and time estimate can be prepared.

*Tool Requirements*

To calculate the tools which we require to carry out the job we need to look at the maximum output we can expect during the contract. This is based on the maximum haulage capacity we can assume. We therefore check from which quarry we will be able to have most lorry trips per day. This in turn will give us the labour requirements for excavation, loading, reshaping, off-loading and spreading, and hence will determine the kind and number of tools we need to have on site.

Quarry III with 14 trips per lorry per day gives us the maximum output. Each day we have to cater for a total of 60 lorry trips (14 trips per lorry multiplied by the average of 4.3 lorries). With 60 trips we will be in a position to transport 240m³ per day and therefore the following number of labourers are required:

| labour requirements for 240m³ | | |
|---|---|---|
| ○ excavation | 240m³/1.2 per work-day | 200 labourers |
| ○ loading | 240m³/7 per work-day | 35 labourers |
| ○ off-load and spread | 240m³/18 per work-day | 16 labourers |
| ○ 10% for miscellaneous + support | | 25 labourers |
| Total | | 276 labourers |

For these 276 labourers we need the following tools (estimate):

| | | | |
|---|---|---|---|
| 100 hoes | @ 10.00 NU | = | 1000.00 |
| 150 pick axes | @ 18.00 NU | = | 2700.00 |
| 200 shovels | @ 13.00 NU | = | 2600.00 |
| 50 mattocks | @ 18.00 NU | = | 900.00 |
| 30 wheelbarrows | @ 62.00 NU | = | 1860.00 |
| 20 bush knives | @ 9.50 NU | = | 190.00 |
| 20 grass slashers | @ 7.00 NU | = | 140.00 |
| 20 rakes | @ 9.0 NU | = | 180.00 |
| 10 spreaders | @ 17.00 NU | = | 170.00 |
| 10 earth rammers | @ 25 NU | = | 250.00 |
| 5 sledge-hammers | @ 22 NU | = | 110.00 |
| 5 crow bars | @ 25 NU | = | 125.00 |
| 5 spirit levels | @ 9.00 NU | = | 45.00 |
| 2 sets of boning rods | @ 6.00 NU | = | 12.00 |
| 10 ranging rods | @ 27 NU | = | 270.00 |
| 4 ditch templates | @ 6 NU | = | 24.00 |
| 2 line and levels | @ 5 NU | = | 10.00 |
| 2 camberboards | @ 10 NU | = | 20.00 |
| strings | | = | 50.00 |
| various small items | | = | 1000.00 |
| total costs for hand tools = | | | 11 656.00 |

We estimate an average life span of 1 year, after which we have to replace the hand tools. This gives us approximately 220 working days. With 276 labourers we can produce (276 labourers x 220 wd) = 60 720 work-days.

The daily rate for hand tools is therefore 11 656.00 NU / 60 720wd = approx. 0.20 NU

Daily labour wage rate: 1.50 NU

Daily gangleader rate: 3.00 NU

As a first step we will need to calculate the direct project cost using the standard calculation chart. To calculate the different BQ items we have to rely on our past experience and our productivity tables (task rates).

We shall include our principal assumptions in the chart, and our calculation chart looks as follows:

| Direct Project Cost Chart | | | | | | | | |
|---|---|---|---|---|---|---|---|---|
| List of quantities taken off drawings | | | | Direct project costs (NU) | | | | |
| Item No. | Description | Unit | Quantity | Labour | Plant Tools | Material | Transp. | Total |
| 1 | Establishment | | | 48.00 | 6.40 | | 576.00 | 630.40 |
| 2 | Improvement of access road | m | 12 000 | 300.00 | 40.00 | | | 340.00 |
| 3 | Quarry preparation | $m^2$ | 5800 | 27.00 | 3.60 | | | 30.60 |
| 4 | Overburden removal | $m^3$ | 1820 | 910.50 | 121.40 | | | 1 031.90 |
| 5 | Reshaping road | m | 15 000 | 562.50 | 75.00 | | | 637.50 |
| 6A | Excavation of hard gravel | $m^3$ | 4180 | 3688.50 | 491.80 | | | 4180.30 |
| 6B | Excavation of very hard gravel | $m^3$ | 7220 | 9025.50 | 1203.40 | | | 10 228.90 |
| 6C | Excavation of soft gravel | $m^3$ | Rate 0.60 | | | | | |
| 7 | Loading | $m^3$ | 11 400 | 2712.00 | 361.60 | | | 3 073.60 |
| 8A | Hauling: average distance 3.530 km | $m^3$ | 4180 | | | | 24 120.00 | 24 120.00 |
| 8B | Hauling: average distance 7.090 km | $m^3$ | 4940 | | | | 42 210.00 | 42 210.00 |
| 8C | Hauling: average distance 1.380 km | $m^3$ | 2280 | | | | 10 854.00 | 10 854.00 |
| 9 | Off-loading and spreading | $m^3$ | 11 400 | 1266.00 | 168.80 | | | 1 434.80 |
| 10 | Watering and compaction | $m^2$ | 60 000 | | | | 17 952.00 | 17 952.00 |
| | Total direct project cost | | | | | | | 116 724.00 |
| 11 | Contingencies (10%) | | | | | | | 11 672.40 |
| | Final total direct project cost | | | | | | | 128 396.40 |

## CALCULATIONS

*Item 1: Establishment*

Assumption: Setting up and removal of 2 camp sites with 8 labourers for 2 days, 2 x 8 x 2 = 32 wd. 4 tractor-days needed for transport.

| Cost: | labour | 32 wd x 1.50 NU = | 48.00 NU |
|---|---|---|---|
| | tools | 32 wd x 0.20 NU = | 6.40 NU |
| | tractor | 4 days x 144.00 NU = | 576.00 NU |

*Item 2: Improvement of access road*

Assumptions: Productivity rate = 60m/wd; total work-days: 12 000m / 60m/wd = 200 wd

| Cost: | labour | 200 wd x 1.50 NU = | 300.00 NU |
|---|---|---|---|
| | tools | 200 wd x 0.20 NU = | 40.00 NU |

*Item 3: Quarry preparation*

Assumption: Average productivity rate = 325m²/wd; total work-days: 5800m² / 325m²/wd = 18 wd

| Cost: | labour | 18 wd x 1.50 NU = | 27.00 NU |
|---|---|---|---|
| | tools | 18 wd x 0.20 NU = | 3.60 NU |

*Item 4: Removal of overburden*

Assumption: Productivity rate = 3m³/wd; total work-days 1820m³ / 3m³/wd = 607wd

| Cost: | labour | 607 wd x 1.50 NU = | 910.50 NU |
|---|---|---|---|
| | tools | 607 wd x 0.20 NU = | 121.40 NU |

*Item 5: Reshaping of road*

Assumption: Productivity rate = 40m/wd; total work-days 15 000 / 40m/wd = 375wd

| Cost: | labour | 375 wd x 1.50 NU = | 562.50 NU |
|---|---|---|---|
| | tools | 375 wd x 0.20 NU = | 75.00 NU |

*Item 6A: Excavation of hard gravel*

Assumptions: Productivity rate = 1.7m³/wd; total work-days: 4180m³ / 1.7m³/wd = 2459 wd

| Cost: | labour | 2459 wd x 1.50 NU = | 3688.50 NU |
|---|---|---|---|
| | tools | 2459 wd x 0.20 NU = | 491.80 NU |

*Item 6B: Excavation of very hard gravel*

Assumptions: Productivity rate = 1.2m³/wd; total work-days: 7220m³ / 1.2m³/wd = 6017 wd

| Cost: | labour | 6017 wd x 1.50 NU = | 9025.50 NU |
|---|---|---|---|
| | tools | 6017 wd x 0.20 NU = | 1203.40 NU |

*Item 6C: Excavation of soft gravel*

Assumptions: Productivity rate = 2.5m³/wd;
Rate: 1.50 NU/day / 2.5m³/wd = 0.6 NU/m³

*Item 7: Loading*

Assumption: Productivity rate = 7m³/wd; total work-days: (11 400m³ / 7 m³/wd) x 1.1 = 1808 wd

| Cost: | labour | 1808 wd x 1.50 NU = | 2712.00 NU |
|---|---|---|---|
| | tools | 1808 wd x 0.20 NU = | 361.60 NU |

Due to the utilization rate used for the lorries (90%) we need to take a lower 'utilization' of labourers into account for activities directly related to the hauling. You would allocate labourers on the assumption that it is possible to achieve the maximum number of trips per day. However, since you calculate with 90% utilization of the lorries you will have a corresponding 'under-utilization' of the labour. This is taken into account by assuming the same 'utilization rate' for the labourers as for the lorries. In practice you multiply the workdays by a factor of 1.11 (100% / 90% = 1.11).

*Item 8A: Hauling average 3.530 km*

Assumptions: Utilization rate 90% (equals calculating with 4.5 lorries instead of 5).

You can make 54 trips per day (12 x 4.5) of 4m³ each;
Total transport capacity: 216m³ per day (54 x 4).
Total number of days needed: 4180m³/ 216m³ = 20 days

The cost per day for lorries is 1206 NU (see 'Equipment requirements')

Cost per day: hired lorries: 50 000 / 66 = 758 NU
our lorries: 448 NU
total: 1206 NU

Cost: transport 20 d x 1206 NU = 24 120.00 NU

*Item 8B: Hauling average 7.090 km*

Assumptions: Utilization rate 90% (equals calculating with 4.5 lorries instead of 5).

You can make 36 trips per day (8 x 4.5) of $4m^3$ each;

Total transport capacity: $144m^3$ per day (36 x $4m^3$).

Total number of days needed: $4940m^3$/ $144m^3$ = 35 days

The cost per day for lorries is 1206 NU (see 'Equipment requirements')

Cost per day: hired lorries: 50 000 / 66 = 758 NU
our lorries: 448 NU

total: 1206 NU

Cost: transport 35 d x 1206 NU = 42 210.00 NU

*Item 8C: Hauling average 1.38 km*

Assumptions: Utilization rate 90%.

You can make 63 trips per day (14 x 4.5) of $4m^3$ each;

Total transport capacity: $252m^3$ per day (63 x 4).

Total number of days needed: 2,280$m^3$/ $252m^3$ = 9 days

The cost per day for lorries is 1206 NU

Cost per day: hired lorries: 50 000 / 66 = 758 NU
our lorries: 448 NU
total: 1206 NU

Cost: transport 9 d x 1206 NU = 10 854.00 NU

It is always good to compare the results of these cost calculations with the estimates done earlier. As an example compare items 8A – 8C with estimates on page 252. Here we arrive at a total cost of 77 184 NU (24 120 + 42 210 + 10 854) for the lorries compared with 79 568 NU on page 252. Both these figures are to a certain extent based on assumptions (e.g. 90% utilization rate) so it is really up to the contractor to decide if the lower figure should be increased before it is included in the bid calculation.

Although we have used the lower figure in the following calculations, this is no indication that it is more correct than the other.

*Item 9: Off-loading and spreading*

Assumption: Productivity rate = $15m^3$/wd; total work-days: (11 $400m^3$ / $15m^3$/wd) x 1.11 = 844 wd

| Cost: | labour | 844 wd x 1.50 NU = | 1266.00 NU |
|---|---|---|---|
| | tools | 844 wd x 0.20 NU = | 169.00 NU |

For comments on 'utilization rates' see Item 7, above.

*Item 10: Watering and compaction*

Assumptions: Tractor with bowser for 66 days @ 144 NU (incl operator) = 9504 NU

2 pedestrian rollers for 66 days @ 128 NU (incl operators) = 8448 NU

Cost: equipment 9504 NU + 8448 NU = 17 952.00 NU

*Preliminaries*

| Item | Description | Cost NU |
|---|---|---|
| P1 | Contractor's site office plus furniture and all temporary buildings | 100 |
| P2 | Supervision:<br>2 Site Supervisors for 4 months (per year 7000 NU) = 2333 NU<br>4 gangleaders for 16 weeks at 5 days = 320 days @ 3.00 NU = 960 NU | 3293 |
| P3 | Land compensation for site camps | 350 |
| P4 | Support at site camp, 7 labourers, 16 weeks of 5 days @ 1.50 NU/d | 840 |
| P5 | Contractor's fully comprehensive insurance | 450 |
| | Total Preliminary Costs | 5033 |

*Risk Allowance*

We are adding the risk allowance as a percentage to the total direct costs. After assessing the risks involved we decide on a risk allowance rate of 3.5% of the total direct costs (116 724 NU) which gives us 4085 NU.

Total risk allowance = 4085 NU

*Company Costs*
To calculate the company costs we have to know the total workload of the company at the time of this particular contract. In our case we have one other small contract and we therefore assume the company costs for the present contract to be 75% of the total costs.

| | |
|---|---|
| Director's salary | 4800 NU |
| Administration | 1800 NU |
| Mechanic | 3500 NU |
| Store Keeper | 1960 NU |
| Bookkeeper, Auditor | 250 NU |
| Interest on bank loan | 450 NU |
| Total for one year | 12 760 NU |

Monthly average costs (12 760/12) 1063 NU

For this contract (1063 x 4 months x 0.75) 3189 NU

The total indirect costs are:

| | |
|---|---|
| Preliminary costs | 5033 NU |
| Risk allowance | 4085 NU |
| Company costs | 3189 NU |
| Total | 12 307 NU |

The total direct costs are 116,724 NU so the indirect costs approximately represent an additional 11%.

Before we add up all the figures we must calculate the expected profit we would like to make. As we think our prices are already on the high side we keep the profit margin at 7% of the direct costs. In our case we shall add the profit as a percentage to each contract item. Our final cost chart now looks like this:

| Item No. | Description | Direct costs NU | Indirect costs 11% NU | Profit 7% NU | Total NU |
|---|---|---|---|---|---|
| 1 | Establishment | 630.40 | 69.34 | 44.13 | 743.87 |
| 2 | Improve access road | 340.00 | 37.40 | 23.80 | 401.20 |
| 3 | Quarry preparation | 30.60 | 3.37 | 2.14 | 36.11 |
| 4 | Overburden removal | 1031.90 | 113.51 | 72.23 | 1217.64 |
| 5 | Reshape road | 637.50 | 70.12 | 44.62 | 752.24 |
| 6A | Excavate hard gravel | 4180.30 | 459.83 | 292.62 | 4932.75 |
| 6B | Excavate very hard gravel | 10 228.90 | 1125.18 | 720.22 | 12 070.10 |
| 6C | Excavate soft gravel (unit rate) | (0.60) | (0.07) | (0.04) | (0.71) |
| 7 | Loading | 3073.60 | 338.10 | 215.15 | 3626.85 |
| 8A | Hauling 3.5km | 24 120.00 | 2653.20 | 1688.40 | 28 461.60 |
| 8B | Hauling 7km | 42 210.00 | 4643.10 | 2954.70 | 49 807.80 |
| 8C | Hauling 1.4km | 10 854.00 | 1193.94 | 759.78 | 12 807.72 |
| 9 | Off-loading and spreading | 1434.80 | 157.83 | 100.44 | 1693.07 |
| 10 | Watering and compaction | 17 952.00 | 1974.72 | 1256.64 | 21 183.36 |
|  | Total Amount | 116 724.00 | 12 839.64 | 8174.87 | 137 738.51 |

As a last step we now have to calculate the unit rates and fill in the BQ. The BQ included as a part of our bid looks like this:

*Bill of Quantities*

| Item | Description | Unit | Quantity | Rate | Amount |
|---|---|---|---|---|---|
| 1 | Establishment | lump sum | | | 744.00 |
| 2 | Improvement of quarry access roads including maintenance throughout the contract period | m | 12 000 | 0.035 | 402.00 |
| 3 | Quarry preparation; consisting of bush clearing, grass cutting, grubbing and clearing well of the quarry | $m^2$ | 5800 | 0.0062 | 36.00 |
| 4 | Excavation of overburden including loading, hauling and stocking within 100m | $m^3$ | 1820 | 0.67 | 1218.00 |
| 5 | Reshaping of road, consisting of re-establishment of carriageway crossfall, reshaping of shoulders and slopes and cleaning of ditches, mitre drains and culverts | m | 15 000 | 0.050 | 753.00 |
| 6A | Excavation of *in situ* gravel and stockpiling ready for loading (hard) | $m^3$ | 4180 | 1.18 | 4933.00 |
| 6B | Excavation of *in situ* gravel and stockpiling ready for loading (very hard) | $m^3$ | 7220 | 1.67 | 12 074.00 |
| 6C | Excavation of *in situ* gravel and stockpiling ready for loading (soft) | $m^3$ | Prov. | 0.71 | |
| 7 | Loading of loose gravel on to trailer or lorry | $m^3$ | 11 400 | 0.32 | 3648.00 |
| 8A | Hauling of gravel to site; average haul distance 3.500km | $m^3$ | 4180 | 6.81 | 28 462.00 |
| 8B | Hauling of gravel to site; average haul distance 7.000km | $m^3$ | 4940 | 10.08 | 49 808.00 |
| 8C | Hauling of gravel to site; average haul distance 1.500km | $m^3$ | 2280 | 5.62 | 12 807.00 |
| 9 | Off-loading and spreading of gravel material to the required thickness | $m^3$ | 11 400 | 0.134 | 1710.00 |
| 10 | Watering and compacting of gravel layer by vibrating roller to the required camber at OMC ±2% in layers of no more than 150mm compacted in carriageway | $m^2$ | 60 000 | 0.36 | 21 183.00 |
| | Total | | | | 137 778.00 |
| | Contingencies, 10% of total | | | | 13 778.00 |
| | Final quotation submitted | | | | 151 556.00 |

The difference between our previous calculation arriving at 137 738 and the figure we have here of 137 778 is very small (40 NU), we can use the figure from the BQ where the rates are slightly rounded off.

Since the contingencies are set aside for extra work currently *not* indicated in the programme it should be clearly separated from the cost for carrying out the project as described. They can only be used following a written confirmation from the client or his or her representative.

However, as the client has asked all bidders to include a post for contingencies amounting to 10% in the total cost for the contract:

The price for the contract to be offered to the client is thus: 151 556 NU

# CHAPTER 10: MANAGING PEOPLE

## Learning Objectives

After you have completed this chapter you should know:

- the basic principles of how to manage a construction business
- how to manage people in labour-based road maintenance work.

## Introduction

Construction work is often divided into technical aspects and managerial aspects. In order to make a success of your contracting business, you and your staff need practical technical experience so that you can complete contract tasks on time and to the correct quality standard. But you will not stay in business very long unless you are also a good manager, so that you plan and complete tasks within cost targets, keep good control of your finances, please your clients and obtain enough work to keep your staff and your equipment regularly employed.

## What is Management?

Management is a complex mixture of art and science; there is not enough space in this chapter to give more than a brief summary of the basic principles to supplement the three IYCB Handbooks and Workbooks. For your purposes as a ROMAR contractor, the task of management can be divided into two parts:

- management of tasks
- management of people.

Earlier chapters in this book have covered the technical and task management aspects of your work as regravelling or routine maintenance contractors, but the management of people has been dealt with only in relation to very specific issues. This does not mean that the management of people is easy, or that you do not need to make an effort in order to get good results. In fact, as a ROMAR contractor using labourers for almost all activities, it will be necessary for you to be an expert in the recruitment and management of large numbers of temporary staff.

It is often said that management of people cannot be learned, because 'good managers are born and not made'. There is some truth in this, and some individuals can build profitable businesses as consultants or traders without employing many people directly. Within the construction industry many builders subcontract most of their work. If you really feel that you will never be able to cope with managing a large labour force you should read no further, because you would be wise to choose a different form of business activity.

This chapter is written for those readers who are still attracted by road maintenance, but who need a checklist of some of the most important things to bear in mind when managing people. In itself, reading this chapter will not make you a good manager. No manager is ever able to achieve perfection, but all managers can gradually improve their performance by reading, thinking, talking to other successful managers, and then trying out new ideas and checking how they work in practice.

*A practical example*

Many books on management theory are difficult to read, because it is hard to relate them to everyday practice. So in order to illustrate management theory in this chapter, we shall relate it to a practical example that follows the theory step by step.

**Your routine maintenance contract**

You have been awarded the contract for the routine maintenance of a 50km gravel road for one year. In view of the need for local employment opportunities, the contract requires you to use labour-based methods whenever possible. You are free to use the lengthman system or the gang system. The only requirement is that a certain quantity of specified maintenance work throughout the entire length of road is carried out every month in accordance with the standards laid down.

You have decided to carry out the work using a gang system. For the entire road length you are employing 33 labourers. You have therefore subdivided the overall length into 3 sections of approximately 16.7km each. For each section a gang of 11 labourers will be responsible for all the routine maintenance activities. Each gang is headed by a gangleader who, at the same time, is one of the 11 labourers.

# Leadership

The best managers are also leaders. It is sometimes said that 'a manager does things right, and a leader does the right thing'. In other words, the manager carries out instructions but the leader also thinks carefully about the task and tries to find the best way of reaching the objective. This means that, as a leader, you should always be looking at how things are done in your firm and asking yourself 'is there a better way?'

You should also remember that the best firms, like the best football sides, are run as teams.

Everyone has some responsibility for management and some responsibility for leadership, because everyone sometimes has to make a decision on how work should be carried out. At the level of the individual labourers the decisions will be relatively simple, such as which tool to use for a particular task. The gangleaders and the site supervisor will have to take more important decisions; and you as the overall manager will have to take complete responsibility for the success of the business.

**Who was to blame?**

Let us assume your gangs are not performing well and the job is done poorly. The client is constantly complaining about it and you are even threatened that your bills will not be paid if the performance does not improve. The client asks you what you think is going wrong. Your answer is: 'My employees are hopeless!'

Now you should ask yourself the following questions:
- who introduced them to the job?
- who trained them on the job?
- who motivated them?
- who plans and monitors their work?
- who organizes their work?
- who is their leader?

**Now who is hopeless?**

# Delegation

As there are so many decisions to be made, the firm will only operate efficiently if there is a system to ensure that each decision is made at the right level. This system will be based on the idea of delegation, which means that you push decision-making down to the lowest level at which the person concerned has the knowledge and facts to make a sensible and quick decision.

## AUTHORITY AND RESPONSIBILITY

As owner/manager you are responsible for the long-term success of your firm and whenever you sign a contract with a client to undertake a project or an annual maintenance contract you accept responsibility for delivering work to the specified standard. As you cannot undertake every operation yourself you now have to delegate responsibility for some subsidiary tasks to other people. This means that you will give them the power to make decisions on your behalf, and they can only do this if you also delegate to them the authority that they need so that the people to whom they give instructions understand that you will back up the decision.

Sometimes the decisions that they make will be different from the ones that you would have made. In such cases it is important that you do not undermine their authority by countermanding their decision or criticizing them in front of other workers. Remember that their mistake is partly your mistake, because you decided that they were fit to make decisions for you. What you should do is discuss their reasons for making the decision, in private, in a quiet and calm way, and explain why you would have acted differently. In such a case your role is that of a management teacher, helping your staff to understand your priorities for the firm and the way in which it conducts its business.

**Delegating authority**

By signing the contract you have accepted responsibility for carrying out the routine maintenance activities along the 50km road. You are delegating the responsibility of carrying out the site activities to your three gangleaders. Each of them is now the manager/leader of a gang of 10 labourers. In order that they can make sure that the work is carried out properly and promptly, you must provide them with the authority that they will need.

## SOURCES OF AUTHORITY

A supervisor's authority comes from three sources:

- conferred authority
- personal qualities
- knowledge and experience.

Conferred authority is given or 'conferred' by an organization, a public authority or the owner of a firm, to recognize a person's job. Sometimes the person concerned has a uniform or a badge of office, like a soldier or a policeman. In other cases, such as your site supervisors, there is no special uniform but all the other workers know that the person concerned has been authorized by you to make day-to-day decisions on the site. They know that the instructions given by your site supervisor will be supported by you, so they will get on with the job.

Many effective managers also gain authority as a result of their personal qualities, which have earned them the respect of the other workers. People who are known to be fair and to keep their word are regarded as having integrity. Integrity is the most important of the personal qualities, and it is often better to choose a site supervisor who has a reputation for integrity than a person who is clever or better educated, but who is thought to be tricky or underhand.

Knowledge and experience are also important sources of authority, particularly in technical areas. Even people who are described as 'unskilled labourers' have in fact very useful knowledge and experience about basic manual tasks. If you are faced with a practical problem you should never be too proud to ask their advice.

**Developing authority**

You should try to develop the authority of your gangleaders in all three areas:

Conferred Authority:

- Define the detailed duties and power of each gangleader.
- Write them down and give a copy to each gangleader.
- Discuss them and make sure they are fully understood.

Personal Qualities:
When you select your gangleaders make sure they have the right personal qualities, and are

- good organizers
- able to deal with the other labourers in a friendly and respectful manner
- reliable and trustworthy
- consistent in applying rules (no favourites; no 'bad days').

Knowledge:

- Select gangleaders who you know have the required practical experience. If you do not have staff of your own from whom you can select, get references for potential staff from their former employers.
- Provide training and make sure they are fully conversant with their job as gangleaders before they take responsibility for work on site. Call them for short training interventions, or provide on-the-job training whenever you think that their performance could be improved.
- Make sure that you always provide the gangleaders with the information they require to carry out their work efficiently. For example, you could arrange brief monthly meetings to inform them about new developments affecting future employment prospects. Give them a chance to ask questions, and answer them as honestly as you can.

# Clear Objectives

In order to be effective a manager at any level must:

- understand the objectives of the work and project
- know the team
- be able to communicate clearly and concisely.

The objectives for a contract manager are to complete the work within the stated targets for:

- quality
- time
- cost.

As well as these overall objectives, there are also a number of contributory objectives that are not explicitly written down in the contract, such as to:

- give the client a good service, which will enhance your reputation
- treat employees fairly
- keep the working sites safe, clean and tidy.

It is necessary that you define your own objectives for each contract and that you make sure that they are clearly understood by all your staff.

**Defining objectives**

Before you sign the contract you should have clearly defined:

- the extent of the routine maintenance activities which your company is committed to carry out (quality, quantity and work methods)
- when you have to carry out these activities (time)
- the unit prices in the contract (cost)
- a policy to give the client the best service you can (reputation)
- how to ensure that your staff are trained and prepared for the job
- plans for efficient organization of the work.

# Asking the Right Questions

It is often said that answering questions is easy; the real skill is asking the right questions at the right time. This is very true in management. There are many things that can go wrong when you are running a business; if marketing is not effective, you will not have enough work; if your financial control is not effective, you could run out of money; if you bid at the wrong price, you will either lose the contract or be faced with carrying it out at unit prices which are not realistic. So as the manager of your business, it is your responsibility to try to guess where the next set of problems might appear and ask the right questions so as to discover the right answers.

**Four questions**

Before you plan your contract you will have to decide how to manage the team of 33 labourers. Here are four sensible questions which will need answers:

- How can I control 33 labourers who are scattered along 50km of roads and make sure they work effectively and please my client?
- What work can I delegate to the three gangleaders and how can I make sure they can control and direct the other labourers efficiently?
- How can I lead 33 labourers who are all different in their individual capabilities, strengths, weaknesses and needs?
- How will I be able to make sure that my labourers perform well, so that at the end of the contract I shall earn the expected profit?

# Know Your Work Team

No two individuals are the same – you have to know your work-team: People have different strengths and weaknesses, needs, personalities, temperaments, backgrounds and ambitions. Effective leaders make it their business to find out these things about the members of their teams.

**Personal contact**

Try to get to know your gangleaders well. This will help you to deal with them on a personal basis, and gain their trust and respect.

Let us suppose that one of the gangleaders is very dedicated and reliable but his educational background makes it difficult for him to fill out the report forms properly. You may then need to take the time to teach him patiently how to fill out reports step by step. On the work site, however, you do not need to control him tightly as you know he is always at the place of duty and makes sure the work is carried out to your full satisfaction.

Another gangleader may be a young woman who has a good educational background, knows her work very well and is ambitious. Unfortunately she is not very tactful, and she might annoy some of the older labourers who may be slow to understand her instructions. In this case you may need to teach her how to approach the labourers in a more tactful way. Explain to her that she should ask you for advice if things go wrong on site, rather than losing her temper and entering into a needless confrontation with her work team.

# Communication

## GENERAL

Good communication is essential in any activity where a lot of people are involved. On a remote site there are many ways in which misunderstandings can arise, and it is your responsibility as the top manager in your firm to ensure that everyone clearly understands their own responsibilities, and the delegated authority which they have in order to carry them out. Communication can be either formal or informal. There is no objection to informal communication in a small organization. In a small firm everybody knows what is going on, and the spoken word is both quick and cheap. However, outside contacts, particularly those with the client, should always be confirmed in writing. In a large firm, or when dealing with staff at a remote site, there is a greater need for written communication, to make sure that the message is received and clearly understood.

Communication in the form of information should always flow in two directions – down and up. The flow downwards from you as the owner/manager will ensure that your staff know enough facts about your business to look after your interests, and also that they will know – rather than trying to guess – the way they should behave when problems arise. The flow upwards will enable *you* to know – rather than guess – the real problems on your sites as well as get an early warning about new business opportunities. Sometimes a response is required, at other times it is merely sufficient for the information to be transmitted and understood.

Information downwards includes:

- answers to questions
- the policy of the company
- objectives for a particular contract
- work plans and targets
- the availability of resources
- feedback on results achieved.

Information upwards includes:

- questions
- suggestions
- information about job problems
- complaints and grievances
- feelings and attitudes
- reports on achieved results.

Sometimes the information upwards will be in the form of bad news, such as an accident or a mistake that could have been avoided. You must always try to remain calm when a subordinate brings bad news, as otherwise they will be encouraged to hide mistakes in future. Construction is a risky business and honest mistakes will always be made. Indeed, it is often said that 'people who never make a mistake never make anything'. The important thing is to learn from mistakes, and you and your staff can only learn and improve if you know about them in the first place.

**Two-way communication**

In your role as the contractor for the maintenance job it is necessary that you communicate well with the field staff and that they communicate back to you and your office.

- Make sure that the three gangleaders get a good introduction to your company and the particular contract, and that they fully understand their role in the work-team.
- Prepare simple work plans for each gangleader and their gang, and set targets which the gangs have to achieve in a particular period of time. Explain the plans and targets carefully to the gangleaders. Also make sure that the gangleaders in turn inform their labourers about the plans.
- Organize regular meetings with your gangleaders at which all aspects of the work can be discussed.
- Pay occasional visits to the field, where you should also talk to the labourers individually and in groups. Give them the chance to ask you questions and make suggestions on the organization of the work.
- Arrange for regular counselling and appraisal sessions with individuals (gangleaders and labourers). For example, you could arrange for such sessions on pay-day when the labourers will visit your office anyway.
- Make sure you have a reporting system in place that allows you to monitor the field activities and the consumption of resources on a daily basis. Each gangleader should complete a daily work report.

## HOW TO COMMUNICATE

When you want to communicate something to someone else, you need to consider:

- whether it needs to be personal or impersonal
- the cost involved
- is feedback required?
- will the message get through?
- accuracy and speed
- is a permanent record required?

You can then decide what means of communication to choose:

- oral communication: you talk to somebody and get immediate answers.
- written communication: this takes time, but has the advantage that it is not easily forgotten. It is especially important when issuing instructions.
- communication by pictures: many subjects can be more easily understood when supported by pictures, sketches, drawings or diagrams, as 'a picture is worth a thousand words'.

**Talking it through**
During the monthly meeting with your gangleaders you tell them that you have observed that the cleaning of ditches is not always done to the correct level. In order to make sure that the gangleaders fully understand your explanations you have prepared a small sketch for each of them, showing the exact requirements.

At the same meeting you inform your gangleaders about a recent increase in the price of cement, and instruct them that they should control the consumption of cement very carefully as the prices for masonry work in the contract cannot be changed. You give this information orally, and your gangleaders discuss with you how cement could be rationed without impairing the quality of work.

**Writing it down**
A gangleader reports that, despite warnings, a certain labourer often reports drunk for duty. He then abuses other team members and performs poorly. As the owner of the company you cannot tolerate such behaviour, and you therefore decide to warn him that he will lose his job if this behaviour continues. In order to make sure that he fully understands your warning and to cover yourself against any possible claims from the local labour office, you send him a warning letter. A copy of the letter will go to his gangleader and another copy to the local labour office to inform them about the incidents and your warning.

# Training Site Supervisors

The daily site supervision will not always be carried out by yourself, but through your site supervisors. These supervisors are the representatives or agents of your company on site and you will have to delegate to them a lot of responsibility. Assume that you are responsible for a regravelling site where you have employed some 100 labourers, 4 tipper trucks for hauling, one tractor with a water bowser, 2 pedestrian vibrating rollers and a site camp with all the tools, material, fuel and other valuables. Besides making sure that these costly goods are properly used and that the labourers are handled correctly you also entrust your site supervisors with the duty of producing quality work at the lowest possible cost. Based on these requirements you will therefore use only the very best people as site supervisors. It is also evident that besides having a sound educational background and practical work experience, continuous training is an essential requirement for a successful performance as supervisor.

In most cases you as the company owner will be directly responsible for the training of your supervisors. If there is no training programme for site supervisors offered by an institution to which you could send your people, you will have to undertake the training yourself. It is therefore essential that you know what and how to train so that the performance of your staff can be sufficiently improved.

In order to carry out this training properly you need to be fully acquainted with the work in all details. However, it is not necessary that you have all the practical skills yourself. You will certainly hire the people who have basic practical skills; what you need to do is to train them based on their existing skills and experience to carry out the job you assign to them. For example, you are employing a foreman who has previously worked for 10 years as a site supervisor on a building construction site. Now you want him to supervise a labour-based road site. It is obvious that he has a lot of experience with construction work which is a very useful basis for his new role. Therefore what you need to do as a first step is to find out what he already knows and what he can do (knowledge and skills). The second step is to identify the knowledge and skills required to carry out the new job on the road site. For example, you find out that he knows how to build retaining walls using rubble stones for houses which are build on a hillside. Therefore you know that he has the basic skills and knowledge to build head and wing walls for culverts on a road. You may only have to explain to him the standard measurements for wing and head walls. He will also be capable of producing the retaining walls of roads in hilly areas. On the other hand, he has never worked on a road site before and therefore he does not even know the standard terminology for road works. If he is to be your efficient road site representative you need to teach him the expressions commonly used so that you can communicate with him on the same level.

To assist you in identifying the knowledge and skills required for site supervisors we provide you with a checklist from which you will be able to prepare your own list. At the same time this list can be used to prepare a job description which you could include in the employment contract for your site supervisors.

*Required knowledge and skills for routine maintenance gangleaders*

Your gangleader must be capable of:

- reading simple written instructions and reporting on the achieved work output
- assessing the daily work requirements and issuing fair task rates according to the instructions
- issuing instructions to the labourers which can be understood by them: which activities, how much, when, and standards to be achieved
- demonstrating to and teaching the labourers how all routine maintenance activities are to be undertaken
- controlling the labourers' work in terms of quality (standards) and correcting where necessary
- assessing the achieved work in terms of quantity and comparing it with the instructions (task rates) given
- communicating effectively with the labourers, the people of the area, and the local leaders and authorities if required.

*Required knowledge and skills for regravelling site supervisors*

*Personnel and administrative activities*

Your site supervisors must know how to:

- prepare daily work plans and fill out site reports
- check the attendance of labourers and record in the muster-roll
- carry out any other site administrative work, as required.

*Technical activities*

Your site supervisors must know how to:

- o organize, direct and control all activities to open quarries, including improvement of quarry access roads
- o organize, direct and control all activities to reshape (rehabilitate) the road before gravelling work can start
- o organize, direct and control all activities to carry out the actual gravelling operations
- o issue appropriate task rates to the labourers for all activities
- o check the performance of all labourers at site and record the achieved work (task rate achieved) in the muster-roll and site reports
- o train (on-the-job), instruct and supervise labourers and gang-leaders how to carry out the gravelling activities.

*Organizational and liaison activities*

Your site supervisors must know how to:

- o plan and authorize the movements of the gravelling equipment according to the company regulations
- o control and record the fuel consumption of the gravelling equipment
- o organize and control the site services and repair work for the gravelling equipment
- o install and maintain the site camps and ensure security
- o control the site store (ordering, storing, recording and issuing).

As you can see, there is a wide range of activities which a regravelling supervisor has to master, and consequently the training cannot be done in a day. A good way to develop your supervisors gradually is to let them go through an on-the-job training programme covering all the activities listed above. During the on-the-job training programme, you have to instruct and demonstrate each activity in turn to your supervisors while performing their duties.

For example, when you introduce your supervisor to your site planning and reporting form you will have to show step by step how to fill in all the details, how to calculate, and what assumptions to make. Then you must let your supervisor try to do it her or himself. In order to make sure everything is done correctly you must check every step afterwards. You will have to continue with this process until you are satisfied with the result. The same applies to all the other activities. For on-the-job training you can use the following teaching pattern:

1 EXPLAIN
2 DEMONSTRATE
3 DISCUSS
4 THE TRAINEES TRY THEMSELVES
5 CORRECT AND DEMONSTRATE AGAIN IF NECESSARY
6 THE TRAINEES TRY AGAIN UNTIL THEY MASTER IT

Where necessary, you also have to provide your staff with simple written instructions to which they can refer in the field. For example, you may produce photocopies of the standards which are attached to the contract document, and explain the details carefully. You can also use other reference material; for example, you can reproduce the routine maintenance activities described in Chapter 7 of this handbook to use as training material for your gangleaders.

It is advisable to organize regular meetings of your site supervisors, which, at the same time, can be used as training sessions. Whenever you visit the site you should note the activities causing problems for your staff. Discuss these problems with them and try to find solutions together with them. If necessary, you may have to provide them with additional instructions and demonstrations.

At the beginning of a job and/or when you have hired new personnel it is very important that you spend enough time training them. The better you instruct your people the more efficient and independent they will operate. This results in a higher profit and gives you time to concentrate on work which only you, as the manager of the company, can do.

# Motivation

Motivation is something within a person which causes them to act. Everyone has certain goals and objectives which they want to achieve. Some are very ambitious, but others just want a living wage and a quiet life. These personal goals and objectives are the engine that drives the many individuals who work for you, and leads them to behave in different ways. It used to be thought that people will only work for money, but most managers now understand that there are a number of other incentives that can motivate staff to work effectively and well. The most important of these is job satisfaction, which can be attained through:

- personal achievement
- recognition
- the work itself
- responsibility
- promotion prospects, and so on.

# Discipline and Morale

Good discipline is created when the members of a team are prepared to observe its rules, even when they have to sacrifice their short-term personal interest in favour of the well-being of the group. Willing self-discipline is always better than that

**Rewarding success**

The road where you are carrying out routine maintenance was previously in poor shape. Since you were awarded the contract half a year ago the road condition has much improved. The local administration has praised your work at the last development committee meeting and during public meetings some of the local leaders have mentioned the good work you are doing. That has given you and your company a reward in terms of official recognition.

You believe that much of the credit is due to the dedication of the gangleader who had previously been involved in building works, but who has made a great effort to understand how to run a road maintenance contract on your behalf. Now you have been awarded a contract for the construction of a large warehouse where you need a reliable foreman. Although it would be easy to advertise for a new foreman and save yourself the trouble of finding a new gangleader, you decide that it is fairer to promote a member of your staff who has made great efforts to support your business. A reliable labourer from the road gang who has shown keen interest in the work and has performed well since the beginning of her employment can now replace the gangleader on the road.

Both of the staff who are promoted will be highly motivated to continue to perform well since their good work in the past has been recognized. Equally important, you are passing a message to all your staff that you know how to reward success and that promotion is based on performance, rather than friendship or family connections.

which is enforced through sanctions. In a minority of cases however, sanctions have to be applied.

The main principles to follow when administering discipline are as follows:

- Make clear to your staff the standards you expect of them.
- If someone misbehaves, warn them clearly of the consequences of their action. Do not make idle threats.
- If you promise disciplinary action in case of a breach, then carry out your promise when the occasion arises.
- Be consistent and impartial at all times.
- Avoid punishing the group unless the group as a whole is at fault. Group punishment upsets the whole group, innocent and guilty alike, and creates opposition to authority.

High morale comes from harmonious and cohesive relationships within a group which has a positive and successful spirit. The willingness of individuals to obey the rules of the group is an important factor. Good morale helps to ensure good discipline.

**Who cares?**

The performance of gang A is better than that of the other two. You decide to find out why, so that you can help the other two gangleaders to learn the 'secret' of gang A's good discipline and high morale. Your investigations produce the following results:

- The gangleader cares about the other labourers.
- All gang members are aware of the work objectives, targets and standards to be achieved.
- All labourers know what is expected of them.
- The labourers are kept informed about the work progress against the set target.
- There is obviously no favouritism.
- The rules of your company are strictly applied.
- The gangleader knows and understands every individual in the gang.
- The gangleader presents any grievances, problems and suggestions promptly and clearly to you as the boss.
- The gangleader defends the group from external criticism, but then acts to eliminate the cause of it.
- The team spirit is encouraged; *WE* are more important than *I*.
- The gangleader continuously demonstrates how the work has to be carried out, and trains the labourers where necessary.

# Incentives

An incentive is offered to get someone to undertake some specific and measurable task. The incentive must motivate people to make an additional effort, over and above what you would expect for their standard daily wage. While motivation is something that drives a person from within, an incentive comes from outside.

Management is about getting things done, with and through people. A manager can motivate people to work harder and more effectively to achieve the goals of the firm by providing the right incentive. Incentives may take many forms. For example, in the case of casual labour the motivation to work comes from the physical need for food and basic commodities. The motivation is strengthened or guided by providing benefits in addition to the basic wage.

Four different kinds of incentives are available:

- threats
- job security
- good leadership
- financial rewards.

Threats and punishment are used to force people to do something they would not otherwise do. This negative incentive has no legitimate place in any kind of commercial company! Job security is a useful incentive, but a contractor is only able to offer job security as long as the contracts keep flowing. Good leadership is a definite incentive, and it pays off in cash terms as well as being the right way to run any kind of business. However, you should also consider the possibility of sharing the benefits of good performance through some form of financial incentive.

## FINANCIAL REWARDS

Financial rewards can be effective incentives in a private company. For casual labourers there are three principal incentive payment systems:

- Daily paid work; in this system a worker is paid a certain amount for every full working day, regardless of what he or she produces. In this case, extra benefits can only be given after normal working hours (overtime).
- Piece-work system; this system of payment is based on the output produced, e.g. for each $m^3$ excavated a certain amount is paid. This allows the workers to earn more than the standard daily pay by producing more. The main advantage of

piece-work is therefore the higher productivity. However, there are several disadvantages: administration is difficult (each individual has a different salary), measuring and control has to be more extensive, work in a team is not easily arranged. For these reasons, piece-work is often very difficult to implement.

- Task work system (sometimes known as 'job and finish'): this system gives the worker one full day's wage for a defined volume of work. The worker is given a task to complete, and is free to leave the site when the task has been approved. This is then counted as a full day's work in the payroll. The task work system has several advantages: planning of the work is simple, because the output is known; one achieves higher output than the daily work, because workers know their task they value the extra free time they get; supervision is easier because each worker knows exactly what, and how much, to do. However, while this arrangement may work well on a job where long-term employment can be guaranteed, on a normal construction site, with many casual workers, the team spirit would not have built up and this system would probably not work.

**Choosing incentives**

For your three gangs you have established the task rate system, because that allows you to plan and supervise the daily work easily and your labourers to attend their farms.

To establish the task rates for the individual activities you have had a meeting with your gangleaders. However, as your gangs work as teams and as it is not very easy to set uniform tasks for all members in routine maintenance, you have decided to let them work on a daily gang task. This means that you have targeted a certain job to be completed by the entire gang on a particular day. You have come to this arrangement because you know that the people of this area would help one another to finish the work, and would not go home early and leave one of their workmates to struggle alone to finish an individual task. You are also aware that it is easier to supervise a gang than 10 individuals. The gang, with its group spirit, itself is taking over part of the supervision and motivation to work well.

However, you have employed the gangleaders on a daily paid work system, as their work cannot easily be measured. Moreover, you have a certain trust in the performance of your gangleaders. During times when emergency maintenance activities are required and when the gangleaders have to attend meetings outside their regular working time you have arranged for an adequate overtime allowance.

## Coping with Complexity

We explained in Chapter 3 that soil is a very variable material and therefore difficult to understand. People are much more variable than soil, so the management of people through the skills of management and leadership is much more complex. In this chapter we have introduced many words and ideas which are relevant to management and leadership:

- DELEGATION
- AUTHORITY
- RESPONSIBILITY
- OBJECTIVES
- COMMUNICATION
- MOTIVATION
- DISCIPLINE
- MORALE
- INCENTIVES

All are important, and all can help you to make your firm stronger and more profitable by turning your work-force into a productive team. Although we have suggested that you should share information – and even some parts of decision-making – with your team, there will remain some things that only you can do. This arises from the fact that only you, as the boss, are in a position to hold the whole firm together, and ensure that it survives to provide continuing employment for all your staff.

# A Helicopter View

In order to achieve this you have to be able to take what has been described as a 'helicopter view' of your business. The helicopter view was identified by management researchers at a major oil company, who were trying to explain why some managers were consistently effective while others always seemed to be running into problems.

What it means is that you have the ability to rise above the hour-to-hour and day-to-day pressures of running your business, to take a broader and more objective view of what is happening to your business and what is likely to happen which will affect its future performance. You can then identify which of your short-term problems need priority attention, and descend 'back to earth' to focus on those which are crucial to the success of the firm.

People who are able to take a helicopter view have an early warning of fundamental changes in their market, and are able to take advantage of them well before their competitors are aware of what is happening. ROMAR is a good example, because it takes advantage of two major changes in the way in which roads are constructed and maintained in low- and middle-income countries:

- The new understanding, by engineers in general, that labour-based methods can produce good results at an economical cost, while meeting the demand of local people for extra employment opportunities.
- The pressure by governments to get more productive and cheaper performance by opening up new opportunities for private contractors to compete for these contracts.

Contractors who rise above their immediate problems to see these trends coming will have learned about the new technology before their competitors, recruited skilled supervisors and obtained the necessary hand tools and basic equipment. In this way, they will have put their firms into the right position to make competitive bids for the new work.

We hope that this ROMAR Handbook and Workbook will help you take advantage of these and other opportunities as they arise.

# ANNEXE

# Labour Standards and Labour-based Road Works

## INTERNATIONAL LABOUR STANDARDS

1. The ILO has developed a system of International Labour Standards which takes the form of Conventions and Recommendations. The Conventions are open to ratification by the member countries. Once ratified, a Convention is binding on that country. Recommendations, on the contrary, set non-binding guidelines to orientate national policy and practice.

2. There are two main groups of International Labour Standards: *basic human rights standards* such as prohibition of forced labour and child labour, freedom of association, equality of opportunity and treatment that can in no way be violated; and *technical standards* that can only be promoted.

3. Six areas of International Labour Standards are of particular relevance to the implementation of labour-based infrastructure work. These are described below, with some comments on their consequences and applicability.

## PARTICULAR STANDARDS

*Forced Labour*

4. There are two Conventions which are relevant in this area. These are the *Forced Labour Convention, 1930 (No. 29)* and the *Abolition of Forced Labour Convention, 1957 (No. 105)*. The first Convention defines forced labour as 'all work which is exacted under the menace of a penalty and for which a person has not offered himself voluntarily.' Certain exceptions applying to labour-based works include minor communal services, some civic obligations and emergency work. The second Convention (No. 105) is supplementary to the first and prohibits the use of forced or compulsory labour in five specific cases. One of these is especially relevant to labour-based road works as it relates to mobilizing non-paid labour for purposes of economic development.

5. An important point in this context is that the continous use of non-paid labour as a self-help contribution is allowed only in

special cases. This is, for example, when a community identifies road improvement and maintenance as a top priority and applies for external assistance (for supervision, construction materials, tools) to complement the labour mobilized for the purpose. In most other cases the use of unpaid labour for road works goes against the spirit of the Forced Labour Conventions. Although this is a Convention that is designed to combat large-scale use of forced labour by governments and large organizations, an important implication of using forced labour is also worth noting for you as a contractor, in addition to all humanitarian considerations. Work carried out by non-paid labour tends to be inefficient because overhead costs for supervision and control are high, and the quality of the work tends to be low.

*Equality of Opportunity and Treatment*

6. In this area, the relevant Conventions are the *Equal Remuneration Convention, 1951 (No. 100)* and the *Discrimination (Employment and Occupation) Convention, 1958 (No. 111)*. The first one lays down the principle of equal pay for men and women for work of equal value. It is important to note that this principle goes beyond equal pay for men and women doing the same job; it means that men and women who do jobs which have the same value – whether the same or not – should be paid equally.

7. The second one (No. 111), which deals with hiring and access to vocational training, prohibits any exclusion, distinction or preference based on race, colour, sex, religion, political opinion, national extraction or social origin. However, to favour especially vulnerable or under-represented groups, through quotas or preferential treatment, so called 'positive discrimination' is not considered to be discrimination.

8. As regards recruitment of a work-force for labour-based work, it is important to note the following: in order to avoid accusations of corruption or favouritism if there are more job seekers than required, which is often the case, one must proceed with a selection process which is generally felt to be fair and not discriminating against any group. A recommended recruitment system which has been tried in several countries is to arrange a secret ballot.

9. The participation of women in labour-based road projects depends largely on the area in which the work takes place, and tends to range between ten and thirty per cent, but cases with up to fifty per cent of women have been recorded. Some projects

set a minimum target for female participation. However, the basic requirement in any case is that women are given a truly equal opportunity to apply for a job and that there is no discrimination in the selection process.

*Freedom of Association*

10. The relevant Conventions here are the *Freedom of Association and Protection of the Right to Organize Convention, 1948 (No. 87)*, the *Right to Organize and Collective Bargaining Convention, 1949 (No. 98)* and the *Rural Workers' Organization Convention, 1975 (No. 141)*. These Conventions cover, respectively: that workers and employers have a right to organize themselves to defend their interests; protection of workers who are organizing themselves, that workers' and employers' organizations do not interfere in each other's affairs, the promotion of voluntary collective bargaining; and the freedom of association for rural workers and encouragement of rural workers to organize and participate in economic and social development.

11. Practical activities in this field include formation of 'Associations of Labour-based Road Contractors' that has taken place in several countries. These associations enable the contractors to better defend their interests. The formation of associations or unions representing important groups of employers or workers often simplifies negotiations on payment, working hours, etc. If such groups are unorganized it would be extremely difficult to reach common agreements which will be respected by all concerned.

*Child Labour*

12. The *Minimum Age Convention, 1973 (No. 138)* states that the minimum age for admission to employment shall not be less then the age of completion of compulsory schooling – normally not less than 15 years. Some countries may, however, specify a minimum age of 14 years. For any type of work that can be dangerous for the health, safety or morals of young persons, the minimum age is usually set at 18 years. For light work, i.e. not physical work, the minimum age is sometimes set at a lower level (13 or 12 years).

*Wages*

13. In this field, two Conventions are most relevant: the *Labour Clauses (Public Contracts) Convention, 1949 (No. 94)* and the *Protection of Wages Convention, 1949 (No. 95)*.

14. The first Convention covers contracts when public funds are awarded by a central public authority to someone employing

workers. Such a contract, for example between a contractor and the Ministry of Works, must contain clauses ensuring that workers' wages, hours of works and other working conditions are at least as good as those stated in the national legislation and collective agreements, and not below the general level in the trade or industry concerned.

15. The second Convention protects workers against excessively low wages by establishing a system of minimum wages. The wage level should take into account the needs (basically food, shelter, clothing) of the workers and their families. Part of the wage may be paid in kind, usually in the form of food rations. This can be a powerful incentive in areas with a shortage of basic consumer goods and high inflation rates. According to a policy agreement between the ILO and the World Food Programme, food rations should not be more than 50 per cent of the total wage to give the workers a right to decide what to spend their money on, and to avoid extensive trading in food rations. The cash portion of the pay should not be less than 50 per cent of the normal wage prevailing in the area or of the applicable minimum wage for the particular work. The payment may, however, be wholly in food in cases where the workers are the direct beneficiaries of the work, for example on self-help schemes. Projects that involve payment in food are usually less suitable for small-scale contractors.

16. Casual workers on labour-based road projects should normally be paid according to the minimum wage legislation in the country. Minimum wages might differ between city and countryside, and according to the category of worker employed. A point to remember is that if you set up a system where you pay your workers according to results, the minimum wage must still be respected. The minimum payment a worker can receive, without obtaining any bonuses must be equal to, or higher than, the minimum wage.

17. When determining the wage rate for casual workers on your road sites there is another issue to consider in addition to wage legislation. It is important that the wages are high enough to attract workers and to ensure satisfactory productivity rates. In employment-intensive works, where all major construction activities are done by labour, the productivity of your labourers is, of course, very important. The labour productivity achieved on your sites usually determines whether you will make a profit or a loss on your contract. Paying reasonably high wages is one very important issue; another is to ensure that workers are paid correctly and on time. Otherwise their working morale will

deteriorate quickly and every dispute will affect your relation with the workers and is likely to cost you a great deal of money.

*Social Security*

18. The *Social Security (Minimum Standards) Convention, 1952 (No. 102)* aims to establish minimum levels of security benefits. It covers nine branches that can constitute a complete 'social security scheme': medical care, sickness, unemployment, old-age, employment injury, family, maternity, invalidity and survivors' benefits. For labour-based road work, the medical care, sickness and employment injury elements are the most relevant.

19. Your permanent staff are often covered by general social security schemes, while the situation of the casual workers hired for the project is different. It is sometimes difficult to extend complete coverage to workers who are temporarily employed. However, you as the contractor are normally obliged, according to the contract, to arrange insurance to cover work accidents and resulting injuries for the entire work-force.

20. It is, of course, very worthwhile and highly recommended to take actions to reduce the risk of injuries; so-called 'preventive measures'. These arrangements normally include: information to workers about hazards, safety training of supervisory staff, provision of protective clothing, use of good quality hand tools and equipment, and arranging safety barriers on dangerous sites.

# References

In preparing the ROMAR Handbook the following references have been consulted.


## CHAPTER 1:

1. *Technical Manual, Volume I, Minor Roads Programme Kenya*, by Intech Associates.
   Published by: Ministry of Public Works, Road Department, Republic of Kenya, January 1992
2. ROMAR Draft Material, ILO Geneva

## CHAPTER 2:

1. *Technology Development in Road Construction and Maintenance*, by R.C. Petts.
   Published by: University of Strathclyde Glasgow, Scotland, Second International Conference on Science and Technology in Third World Development, April 1993
2. *Management and Supervision of Labour Based Road Construction and Maintenance, Training Course Notes for Technical Staff of the District Council Road Units, Republic of Botswana*, by Intech Beusch & Co.
   Published by: Roads Training Centre, Ministry of Works, Transport and Communications, Republic of Botswana, August 1992

## CHAPTER 3:

1. *Guide to the training of supervisors for labour-based construction and maintenance, Vols. I and II: Trainees' Manual, Vol. III: Instructor's Manual; International Labour Office*, by L.S. Karlsson and J.J. de Veen.
   Published by: International Labour Office CH-1211 Geneva, 1981
2. *Rural Transport Services, a guide to their planning and implementation*, by Henry L. Beenhakker with S. Carapetis, L. Crowther and S. Hertel
   Published by: Intermediate Technology Publications, 103–105 Southampton Row, London WC1B 4HH, UK, 1987

## CHAPTER 4:

1. *International Course for Engineers and Managers of Labour-based Road Construction and Maintenance Programmes*, by Andreas Beusch and Jan de Veen.
   Published by: International Labour Office CH-1211 Geneva, 1991
2. *LCU Training Material*, by Htun Hlaing
   Published by: LCU, Maseru
3. *Equipment Maintenance*, by Jim Hamper.
   Published by: International Labour Office CH-1211 Geneva, Report of Proceedings – Labour-Based Technology, A Review of Current Practice (CTP 133), Seminar October 1993, Zimbabwe Institution of Engineers, October 1993

## CHAPTER 5:

1. *International Course for Engineers and Managers of Labour-based Road Construction and Maintenance Programmes*, by Andreas Beusch and Jan de Veen.
   Published by: International Labour Office CH-1211 Geneva, 1991
2. *Guide to the training of supervisors for labour-based construction and maintenance, Vols. I and II: Trainees' Manual, Vol. III: Instructor's Manual; International Labour Office*, by L.S. Karlsson and J.J. de Veen.
   Published by: International Labour Office CH-1211 Geneva, 1981
3. *Technical Manual, Volume I, Minor Roads Programme Kenya*, by Intech Associates.
   Published by: Ministry of Public Works, Road Department, Republic of Kenya, January 1992
4. *Management and Supervision of Labour Based Road Construction and Maintenance, Training Course Notes for Technical Staff of the District Council Road Units, Republic of Botswana*, by Intech Beusch & Co.
   Published by: Roads Training Centre, Ministry of Works, Transport and Communications, Republic of Botswana, August 1992
5. *Roads 2000 Works Manual (Draft)*, Ministry of Public Works, Roads Department, Republic of Kenya, Intech Associates, 1994.
   Published by: (not yet published)

## CHAPTER 6:

1. *Maintenance Manual, Minor Roads Programme Kenya*, by Intech Associates.
   Published by: Ministry of Public Works, Road Department, Republic of Kenya, November 1992.
2. *Management and Supervision of Labour Based Road Construction and Maintenance, Training Course Notes for Technical Staff of the District Council Road Units, Republic of Botswana*; by Intech Beusch & Co.
   Published by: Roads Training Centre, Ministry of Works, Transport and Communications, Republic of Botswana, August 1992
3. *International Road Maintenance Handbook, Volumes I to IV (PIARC Road Maintenance Handbook)*, by PIARC , revised R.C. Petts
   Published by: Transport Research Laboratory, UK, 1994 (English Version)

## CHAPTER 7:

1. *Maintenance Manual, Minor Roads Programme Kenya*, by Intech Associates.
   Published by: Ministry of Public Works, Road Department, Republic of Kenya, November 1992.
2. *Management and Supervision of Labour Based Road Construction and Maintenance, Training Course Notes for Technical Staff of the District Council Road Units, Republic of Botswana*, by Intech Beusch & Co.
   Published by: Roads Training Centre, Ministry of Works, Transport and Communications, Republic of Botswana, August 1992
3. *Roads 2000 Works Manual (Draft)*, Ministry of Public Works, Roads Department, Republic of Kenya, Intech Associates, 1994.
   Published by: (not yet published)
4. *Labour-Based Contract Maintenance Programme, Orientation Course for District Engineers, Ministry of Works, Transport and Communications, Republic of Uganda*, by Intech Beusch & Co.
   Published by: International Labour Office CH-1211 Geneva, April 1993

## CHAPTER 8:

1. *Technical Manual, Volume I, Minor Roads Programme Kenya*, by Intech Associates.
   Published by: Ministry of Public Works, Road Department, Republic of Kenya, January 1992
2. *International Course for Engineers and Managers of Labour-Based Road Construction and Maintenance Programmes*, by Andreas Beusch and Jan de Veen.
   Published by: International Labour Office CH-1211 Geneva, 1991
3. *LCU Training Material*, by Htun Hlaing
   Published by: LCU, Maseru

## CHAPTER 9:

1. *IYCB Handbook 1, Pricing and Bidding*, by Claes-Axel Andersson, Derek Miles, Richard Neale and John Ward.
   Published by: International Labour Office CH-1211 Geneva, 1994

## CHAPTER 10:

1. *International Course for Engineers and Managers of Labour-based Road Construction and Maintenance Programmes, Management of People and Management Self-Development*, by J. Varjo
   Published by: International Labour Office CH-1211 Geneva, October 1993
2. *Management and Supervision of Labour-Intensive Road Construction and Maintenance, Training Course Notes for Technical staff of the District Council Road Units, Republic of Botswana*, by Arne Engdahl
   Published by: Roads Training Centre, Ministry of Works, Transport and Communications, Republic of Botswana, January 1988
3. *IYCB Handbook 2, Site Management*, by Claes-Axel Andersson, Derek Miles, Richard Neale and John Ward.
   Published by: International Labour Office CH-1211 Geneva, 1996

# Recommended Reading

This is a list of recommended titles for further detailed reference on specific topics to supplement the ROMAR modules.

## MAINTENANCE

*International Road Maintenance Handbook, Volumes I to IV (PIARC Road Maintenance Handbook), by PIARC, revised R.C. Petts.*
Published by: Transport Research Laboratory, UK, 1994 (English Version)
*may be ordered from:* Transport Research Laboratory
Crowthorne, Berkshire RG11 6AU
United Kingdom
(free issue for recipients in developing countries)

## CONSTRUCTION

*An introduction to labour-based road construction*, prepared for the International Labour Office by J. Antoniou, P. Guthrie and J. de Veen.
Published by: Longman Group (UK) Ltd, 1990, ISBN 0 582 06060 5
*may be ordered from:* ILO Geneva, or any bookshop

## CONTRACT MANAGEMENT

*Improve Your Construction Business (IYCB), (Handbook/Workbook 1: Pricing and Bidding; Handbook/Workbook 2: Site Management; Handbook/Workbook 3: Business Management)*, by Claes-Axel Andersson, Derek Miles, Richard Neale, and John Ward.
Published by: ILO, Geneva, 1994–95
*may be ordered from:* ILO Geneva, or any bookshop

*Interactive Contractor Training (Module 1: Estimating and tendering; Module 2: Project Planning; Module 3: Site Productivity)*, by Tor Hernes, edited by Derek Miles.
Published by: ILO, Geneva, 1988
*may be ordered from:* ILO Geneva, or any bookshop

www.ingramcontent.com/pod-product-compliance
Lightning Source LLC
Jackson TN
JSHW011350130125
77033JS00015B/548